Industrial Materials

Larry D. Helsel
Eastern Illinois University

Peter P. Liu
Eastern Illinois University

D1636863

Publishers
The Goodheart-Willcox Company, Inc.
Tinley Park, Illinois

Library of Congress Card Catalog Number 00-052789
International Standard Book Number 1-56637-815-X

1 2 3 4 5 6 7 8 9 10   01   05 04 03 02 01

Library of Congress Cataloging-in-Publication Data

Helsel,Larry David,
    Industrial materials / by Larry D. Helsel, Peter P. Liu.
        p. cm.
    Includes index.
    ISBN 1-56637-815-X
    1. Materials. I. Liu, Peter P., II. Title
TA404.8 .H45 2001
620.1'1--dc21   00-052789
                                                                  CIP

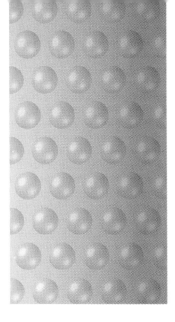

Introduction

The last half of the twentieth century experienced a revolution in materials science and technology that parallels, if not exceeds, the automation and manufacturing processes revolution of the early part of the twentieth century. Although materials science and technology has traditionally taken a backseat to manufacturing processes, its importance to contemporary industry cannot be overstated. Materials account for as much as 30 to 50 percent of the cost of manufactured goods. Extensive research and development efforts are devoted to reduce the overall material cost while increasing the quality and performance of manufactured products.

The service requirements of high performance space and military systems led to the development of advanced ceramics and composites. These materials, first developed for the military, are now reaching the commercial markets and their use is projected to increase dramatically during the next decade. Markets for advanced ceramics grew from less than $2 billion in 1987 to more than $20 billion in 2000. New applications are continuously being proposed for these materials, requiring more extensive research and development of processes to satisfy the demands of high volume industries. The sporting goods, automotive, and commercial aircraft industries have taken the lead in using advanced engineered materials. As new, more efficient manufacturing methods are developed for these materials, their use will increase. According to a report published by the United States Office of Technology Assessment, a key to remaining competitive with other countries and companies is to educate and train more scientists and technologists with a broad background in advanced materials. The purpose of this book is to provide a strong foundation of knowledge of industrial materials, ranging from traditional metals, ceramics, and polymers to advanced engineered materials and composites.

In addition to providing current knowledge of conventional and advanced materials, this text includes discussions of standards and standards organizations, properties and nature of materials, materials testing, and applications. Also, Internet sites related to the discussion are provided at the end of each chapter. The authors made every attempt to write this book in a manner which makes the science and technology of materials easy to understand for all students.

This book is suited for students in such disciplines as engineering, engineering technology, and industrial technology, either in two-year or four-year programs. Readers may find this book useful for other scientific or technical programs as well.

About the Authors

Dr. Larry D. Helsel is a professor in the School of Technology at Eastern Illinois University. Dr. Helsel has more than twenty years of experience in higher education, with most of his teaching experience in the areas of materials science, materials testing, metallurgy, and manufacturing processes. Dr. Helsel received his bachelor's and master's degree from the California University of Pennsylvania and his doctorate from the Pennsylvania State University. In addition to numerous journal articles and professional presentations, Dr. Helsel is coauthor of a textbook on materials and processes. He has been a member of the National Association of Industrial Technology for more than twenty years and has served in several leadership roles with NAIT. Dr. Helsel has completed a number of postdoctoral workshops and courses in computer-integrated manufacturing, quality assurance, methods analysis, computer simulation, industrial engineering, and computer control programming.

Dr. Peter Ping Liu obtained his B.S. degree in electrical and mechanical engineering from Nanchang University, China, in 1982. He earned an M.S. degree in materials science and engineering from Zhejiang University, China, in 1984, and completed a Ph.D. degree in mechanical engineering from Iowa State University in 1991. His major research interests rest on biomedical polymers, polymer recycling, polymer rheology, materials tribology, electrical sliding, corrosion, and failure analysis. He is a registered professional engineer (PE) in the state of Illinois, a certified quality engineer (CQE) by American Society for Quality, and a certified senior industrial technologist (CSIT) by National Association of Industrial Technology. He is currently a full professor at Eastern Illinois University.

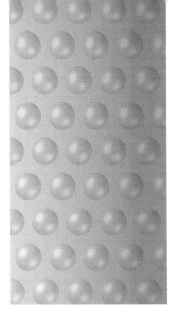

Table of Contents

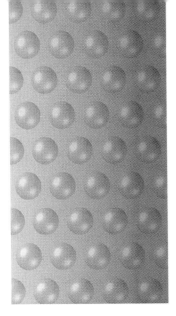

Chapter 1

Introduction to Industrial Materials

KEY CONCEPTS

Upon completion of this chapter, you should understand:

> The basic historical use of materials by humans to extend human potential.

> The importance of using natural resources wisely and the consequences of acting responsibly to recycle materials.

> The process used to design new products or redesign existing ones.

> The concept of concurrent engineering of product and process.

> The definitions and importance of specifications and standards.

> The purposes and functions of standards organizations and trade associations.

HISTORY

Materials have been shaped or used as found in nature to extend human potential since before recorded history. Early humans used rocks shaped by the forces of nature for hunting and protection millions of years ago, Figure 1-1. These *eoliths,* as they are called by archaeologists, were generally found in creek or riverbeds where they were shaped by water, sand, and wind over millions of years.

Figure 1-1.
Early humans shaped stone tools for specific purposes.

Slowly, humans learned to shape stone for specific purposes by striking them together. Hand axes, scrapers, and other tools and weapons have been found in ancient civilizations dating back more than 100,000 years. Later, the discovery of fire led to the use of materials such as ceramics from which cooking utensils and storage vessels were made.

Some archaeologists believe that fire also led to the discovery and use of pure metals. Copper, gold, silver, and meteoric iron were the earliest metals used because they were easily found in nature. The discovery and use of these metals eventually led to the development of mining and the discovery that metals could be alloyed (mixed with one another) to produce other materials.

Human invention, innovation, and discovery from these early beginnings have led to a world of materials technology that allows humans to explore the universe and continuously improve our standard of living.

MATERIALS AND THE ENVIRONMENT

All of this advancement is not without cost, however. Since the Industrial Revolution (circa 1760-1840), humans have used more and more materials from nonrenewable sources. *Nonrenewable resources* are resources that cannot be quickly replaced by natural ecological cycles. These include fossil fuels and inorganic minerals, such as iron. Nonrenewable resources are finite. They will not last forever. In fact, some predict that many nonrenewable resources will be depleted by the middle of the 21st century. Mineral ores for copper and iron that were once 60–90% pure are now as little as 4–6% pure. More energy is needed to refine these low grade ores and more waste is generated.

Renewable resources are those that can be quickly replaced by natural ecological cycles or sound management practices. These include wood products; natural fibers, such as cotton and silk; and polymers made from plants, such as natural rubber.

Scientists often discover after the fact that many manufactured materials are harmful to humans and to the environment. Some processes used to refine the raw materials produce byproducts found to be harmful to animals, humans, and the environment.

The scientists and technologists that have given us throwaway containers, disposable diapers, inexpensive appliances, and cars are now realizing that the "throwaway society" is facing one of its greatest challenges ever. For example, Americans generate an average of more than 1/2 ton of garbage per person per year. In their lifetimes, Americans will throw away 600 times their adult weight in garbage. More and more countries are developing the technologies that require the use of nonrenewable resources. Therefore, the production of pollution and solid waste is not restricted to the leading industrialized nations anymore.

Environmentalists are discovering the dangerous consequences of throwing materials and chemicals into landfills. Poisons are running off landfills into water supplies. Also, landfills are rapidly being filled to capacity. Seventy percent of America's 20,000 landfills closed between 1978 and 1988. Approximately 2000 more closed by 1993. The Environmental

Protection Agency (EPA) estimated the total municipal solid waste in the US to be 223 million tons in the year 2000 with an annual growth rate of 1.2% per year. At this rate, the total municipal solid waste will top 262 million tons by 2010.

Many of the materials discarded in landfills can be recycled to make new products. Significant gains are being made in recycling. Recycling not only reduces the amount of nonrenewable resources used and the volume of garbage going into landfills, but can also reduce the amount of energy required for refining materials.

For example, making one ton of paper from recycled material uses only about 60% of the energy needed to make virgin paper. Recycling glass can save approximately 30% of the energy needed to make glass from raw materials. Mining waste is also significantly reduced. Glass represents about 8% of the waste in landfills in the United States.

Tremendous amounts of electricity are needed to electrolytically remove metallic aluminum from aluminum oxide ore. Remelting recycled aluminum takes only a fraction of the energy needed to refine it from the ores that contain aluminum.

Much of the gains in recycling are a result of private citizen's groups concerned about the environment. Corporations are also becoming more and more environmentally conscious. Aluminum companies, recognizing the economic advantage of recycling, initiated recycling programs that have resulted in approximately 60% of the aluminum cans being recycled.

In too many cases, however, recycling provides little or no economic advantage. For many companies, the economic disadvantage of recycling outweighs the motivation to save the environment. Every day Americans throw away about 44 million newspapers, or the equivalent of 500,000 trees per week. Experts estimate that paper constitutes between 40–50% of the waste stream. Recycling paper would reduce energy consumption, save trees, and significantly reduce the volume of garbage taken to landfills. Unfortunately the economic incentive is not there to aggressively recycle paper.

Materials technologists, engineers, and scientists must understand both the short- and long-term consequences of their decisions. The criteria for selecting materials and processes must extend beyond economics. They must understand the consequences of making certain decisions. Their study must be broader than the properties, classifications, and characteristics of metals and nonmetals.

THE DESIGN PROCESS

Product design or redesign is initiated as a result of general market demands or trends, customer requests, or new ideas. For example, the fuel crisis of the 1970s created a market demand for smaller, lighter, fuel-efficient cars. Research and development may create new materials for which designers and engineers must find new applications.

Science and technology, consumer spending patterns, economics, general demographics (age, gender, ethnicity), and competition all influence the design process and selection of materials in a significant way. The key to being successful in today's market is to shorten the time from concept to production as much as possible. To accomplish this effectively requires a team effort. The design team in today's highly competitive market often consists of product design engineers, process engineers, manufacturing engineers, marketing specialists, and other business management team members.

This approach, known as ***concurrent engineering***, is much different than the traditional approach. The traditional approach relies on the design engineer to develop the product design without input from process and manufacturing engineering. Concurrent engineering develops products and their related manufacturing and support processes at the same time. In concurrent engineering, decisions about materials, manufacturing processes, assembly methods, and costs are made by team consensus.

Designing the product and process concurrently shortens the time from concept to production. This reduced ***time to market*** is critical in today's worldwide competitive market. The ability to design and simulate performance on computers also contributes to reduced time to market. Computers also allow the design teams to quickly build prototypes. ***Rapid prototyping*** uses a technology known as stereolithography to produce a prototype in a very short period of time. This gives the design team an enormous advantage over traditional methods of prototype modeling.

Stereolithography is a process that is used to create three dimensional plastic prototypes from CAD/CAM/CAE data, Figure 1-2. This acronym stands for Computer-Aided Drafting/Computer-Aided Manufacturing/Computer-Aided Engineering. Stereolithography uses laser, optical scanning, chemistry, and computer technologies. In this process, CAD/CAM/CAE data

Figure 1-2.
This rapid prototyping machine can produce sample parts based on computer designs. (Stratasys, Inc.)

are used to generate a series of cross sectional profiles of the design. A laser generating an ultraviolet (UV) beam is moved across the surface of a vat of liquid photopolymer by a computer-controlled scanning system. The laser changes the liquid into a solid as the computer draws each cross section. A vertical elevator system lowers the newly formed layer. Successive cross sections are created until the complete model is formed.

Advanced computer workstations are now being used for three-dimensional design and analysis to model and simulate product and process reliability. These tools and other advanced technologies have greatly improved the design and manufacturing process.

MATERIAL SELECTION

The selection of the materials used in the product design is as important as any other phase of development. Materials can represent 30–50% of the cost of the product. A critical error in material selection can affect every area of manufacturing and marketing. For example, if the cost of raw materials is the only criterion used, the manufacturing processes may be so expensive that the product is unprofitable.

Matching materials to product and process requirements is a critically important task for the design team. The team must analyze customer requirements and prepare product specifications based on the operational conditions proposed for the product. The *operational conditions* are the conditions in which the product is expected to perform while in service. For example, the product may be expected to operate in extreme heat, high humidity, and in contact with an oil spray. These are environmental conditions. The operational conditions may also require resistance to mechanical forces such as tension, compression, fatigue, or others.

The material selected must have the desired properties to ensure that the product will be reliable. *Reliability* is the degree of probability that a product will function as specified in service for the intended life of the product without failure. The following factors are generally considered in selecting materials for a product.

◆ Properties
◆ Cost
◆ Source
◆ Environmental Impact

PROPERTIES

The properties important to the selection process are classified as mechanical, physical, or chemical. *Mechanical properties* include hardness, tensile strength, compression strength, wear, stiffness, shear, impact, and others. *Physical properties* relate to the characteristics of the material when it interacts with different forms of energy, such as light, heat, electricity, and magnetism. *Chemical properties* include composition, corrosion resistance, and flammability.

The mechanical, physical, and chemical properties of a material must meet the operational conditions as well as the reliability requirements for the service life of the product. For example, selecting a material for its mechanical properties without considering its resistance to corrosion could have serious consequences and may result in serious injury or loss of life.

COST

The economics of the material selection is usually of primary importance. Companies are in business to make a profit and must always consider cost when selecting materials. Several factors must be considered in analyzing the cost of materials. The cost per unit of the raw material is only part of the analysis. Other factors to consider are:

- ◆ The cost of processing
- ◆ The cost of shipping
- ◆ The cost of storing
- ◆ The cost of disposing of waste

Purchasing a material simply because of its initial low raw material cost may not be economically sound when other costs are considered. For example, a material may be environmentally safe and meet all mechanical, physical, and chemical property requirements but be extremely difficult and costly to fabricate.

The Cost of Storing

Some materials require special conditions for storage. An environmentally controlled climate may be required in order to preserve the material until used in production. Some materials have a *shelf life.* This means that they must be used within a specified period of time or they will not perform as specified. For example, polystyrene beads used in the production of many foamed polystyrene products generally have a shelf life of less than one year.

The Cost of Waste Disposal

The disposal cost of waste resulting from defective parts or simply from the production process is an increasingly large problem. The methods of disposal are becoming tightly controlled due to a greater concern by all segments of society for the environment. The Environmental Protection Agency (EPA) has established policies based on federal law that tightly control the disposal of many industrial materials.

Some materials can be recycled from waste or defective products to make new parts. Other waste simply must be thrown away. For example, thermoplastics, a classification of polymers, are molecular materials which can be recycled by reprocessing. Thermoset plastics, however, have crosslinked polymer chains that prevent them from being recycled. If the thermoplastic meets all other requirements, it may be the correct choice because of its ability to be recycled.

SOURCE

The source of the material is another factor that must be considered when selecting a material. The material must be available in sufficient quantities and be delivered in a timely manner to meet production schedules. Production delays and alternate sources can be costly. If the source is unreliable, the ultimate cost may be much higher than expected.

ENVIRONMENTAL IMPACT

More and more designers and corporate managers are facing the need to consider environmental impact when selecting materials and processes. In some cases, the motivation comes from their own civic sense of doing what is right for society. In other cases, the legal requirements and costs associated with protecting the environment are motivating factors.

While plastics have become the most popular choice of material because of their low cost and ease of manufacturing, many of the chemicals used to manufacture them are at the top of the Environmental Protection Agency's list of most hazardous materials. For example, the chemical propylene, used to make a plastic called polypropylene, is one of the worst toxins according to the EPA. When polypropylene is incinerated, it gives off nickel. Nickel is a toxic metal which may cause certain forms of cancer.

Another problem with some materials is biodegradability, or the ability to decompose. The controversy surrounding products made from foamed polystyrene is a good example. Polystyrene products are completely *nonbiodegradable*. This means that they do not easily decompose. When disposed of in landfills, they will still be there hundreds of years from now. When disposed of in the oceans, polystyrene has been found to be harmful to marine life. Many fast-food restaurants have greatly reduced or eliminated foamed polystyrene products because of the controversy.

SPECIFICATIONS AND STANDARDS

In the process of selecting materials for a product, the design team must consider the specifications for the product. Specifications usually state requirements for strength, mechanical, physical, chemical, and dimensional properties. They may also state requirements for quality and the material's ease of fabrication, Figure 1-3. Refer to the following definition of specifications.

> *Specifications:* Clear and accurate descriptions of technical requirements of materials, products, or services. They may state requirements for quality and the use of materials and methods to produce a desired product, system, application, or finish.

Designers rely on established standards and organizations that govern standards when determining what materials meet the requirements of the operational conditions and desired quality. Standards are widely accepted or acknowledged measures of comparison for quantitative or qualitative values. Refer to the following definition of standards.

> *Standards:* Specifications that have been adopted for use by a broad group of manufacturers, users, or specifiers.

Standards have been established for the conversion of raw materials to commonly accepted shapes and sizes. For example, iron ore and the other raw materials used in iron and steel production are refined into billets, sheets, slabs, and structural members of standard size and shape, Figure 1-4. Standards have also been established for the application of these standard materials. For example, there is a standard established for the maximum horizontal span of a steel I beam for a given load.

Figure 1-3.
Customer specifications are
often provided in the form
of a drawing.

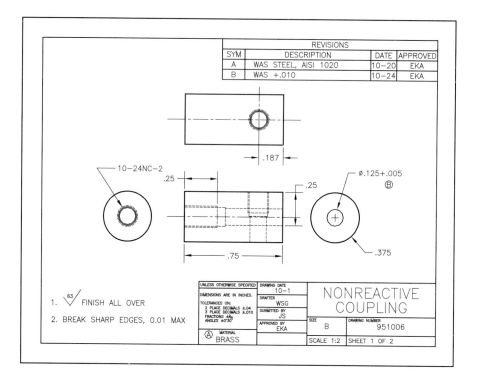

STANDARDS ORGANIZATIONS

There are several private and public organizations that establish standards for comparison of qualitative and quantitative values. Private-sector organizations that establish standards for materials and processes are generally known as trade associations. *Trade associations* are organizations of individual manufacturers or businesses engaged in the production or supply of materials and/or services of a similar nature. The membership list of the Asphalt Institute, for example, includes nearly every petroleum refining company in the United States and abroad.

One of the most important activities of a trade association is the conduct of research related to the improvement of their materials and methods. This research helps formulate performance standards. Some common trade associations are:

♦ American Iron and Steel Institute (AISI)
♦ The Engineered Wood Association (APA)
♦ Society of Plastics Engineers (SPE)
♦ National Association of Home Builders (NAHB)
♦ Asphalt Institute
♦ Society of Automotive Engineers (SAE)
♦ American Society for Metals (ASM)

This is only a very small listing of the hundreds of trade associations that establish performance standards. Their initials and emblems can usually be found on the products for which they establish standards. Look closely the next time you purchase oil for your car or plywood for a building project and you will find the trade associations emblem, Figure 1-5.

Figure 1-4.
Raw steel is processed into billets, sheets, and structural shapes. (American Iron and Steel Institute)

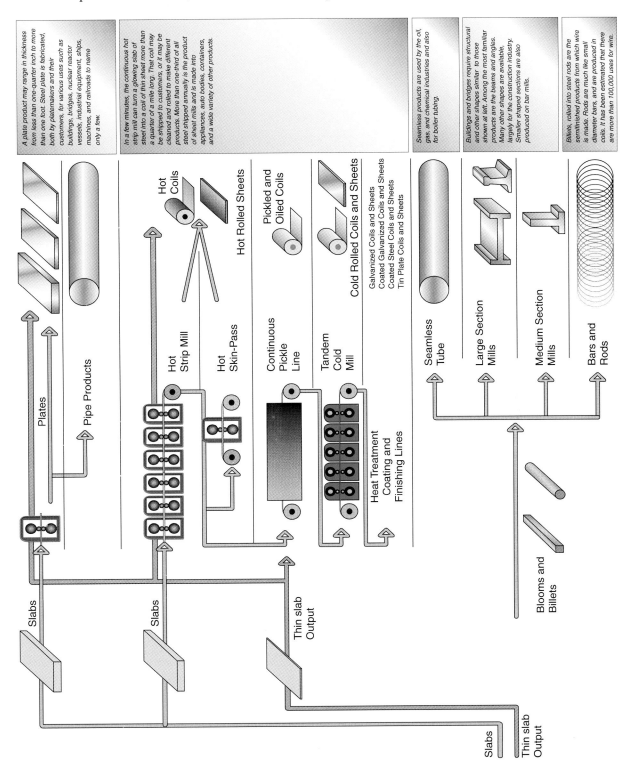

A plate product may range in thickness from less than one-quarter inch to more than one foot. Steel plate is fabricated, both by platemakers and their customers, for various uses such as buildings, bridges, nuclear reactor vessels, industrial equipment, ships, machines, and railroads to name only a few.

In a few minutes, the continuous hot strip mill can turn a glowing slab of steel into a coil of thin sheet more than a quarter of a mile long. That coil may be shipped to customers, or it may be cleaned and rolled to make different products. More than one-third of all steel shipped annually is the product of sheet mills and is made into appliances, auto bodies, containers, and a wide variety of other products.

Seamless products are used by the oil, gas, and chemical industries and also for boiler tubing.

Buildings and bridges require structural and other shapes similar to those shown at left. Among the most familiar products are the beams and angles. Many other shapes are available, largely for the construction industry. Smaller shaped sections are also produced on bar mills.

Billets, rolled into steel rods are the semifinished products from which wire is made. Rods are much like small diameter bars, and are produced in coils. It has been estimated that there are more than 100,000 uses for wire.

Plates

Pipe Products

Hot Coils

Hot Rolled Sheets

Pickled and Oiled Coils

Cold Rolled Coils and Sheets

Galvanized Coils and Sheets
Coated Galvanized Coils and Sheets
Coated Steel Coils and Sheets
Tin Plate Coils and Sheets

Hot Strip Mill

Hot Skin-Pass

Continuous Pickle Line

Tandem Cold Mill

Heat Treatment Coating and Finishing Lines

Seamless Tube

Large Section Mills

Medium Section Mills

Bars and Rods

Slabs

Slabs

Thin slab Output

Blooms and Billets

Slabs

Thin slab Output

Figure 1-5.
This trade association stamp from The Engineered Wood Assoication (APA) is found on all plywood panels. (The Engineered Wood Association (APA))

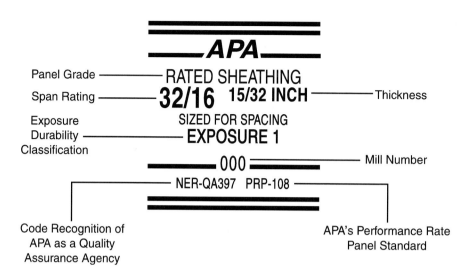

In addition to trade associations, there are several private and government-sponsored organizations that set performance standards. Perhaps the best known of these are:

♦ American Society for Testing and Materials (ASTM)
♦ American National Standards Institute (ANSI)
♦ National Institute for Standards and Technology (NIST)
♦ International Organization for Standardization (ISO)

THE AMERICAN SOCIETY FOR TESTING AND MATERIALS (ASTM)

The American Society for Testing and Materials (ASTM) is an international, privately financed, nonprofit, technical, scientific, educational organization. The society is dedicated to "the promotion of knowledge of the materials of engineering, and the standardization of specifications and methods of testing." ASTM includes approximately 32,000 individual members and 4,000 organizational members from more than 100 countries. Standards are formulated and recommended by 129 technical committees and 2080 subcommittees.

ASTM publishes its standards in a multivolume set titled *ASTM Standards,* Figure 1-6. Contained within these volumes are the definitions of terms, specifications, and methods of testing for a wide variety of materials. ASTM specifications are designated by an alphanumeric code. The last portion of the code is the year of adoption or revision. For example, C150-89 is the standard specification for portland cement. This standard was last revised in 1989.

Figure 1-6.
The American Society for Testing and Materials produces volumes of standards related to testing of materials.

THE AMERICAN NATIONAL STANDARDS INSTITUTE (ANSI)

ANSI is the name adopted by the United States of America Standards Institute in 1966. ANSI is made up of the Member Body Council and the Company Member Council. The Member Body Council members consist of national trade, professional, and scientific associations. The Company Member Council consists of representatives from labor, government, and several hundred industrial firms. These two councils maintain procedures for approval of standards, recommend areas of standardization considered essential, and review standards. ANSI also considers promoting education as an important objective of their organization. Approximately 3,000 standards have been developed and approved by ANSI.

THE NATIONAL INSTITUTE FOR STANDARDS AND TECHNOLOGY (NIST)

The National Institute for Standards and Technology (NIST) is a federal agency under the direction of the Department of Commerce. In the past, this agency was called the National Bureau of Standards. NIST is a center for science and engineering research. NIST is a valuable partner of industry in providing the standards and measurement techniques that further technological advancement in domestic and international commerce. With the enactment of the Omnibus Trade and Competitiveness Act of 1988, NIST was designated the federal government's catalyst for speeding innovation and accelerating the adoption of new technologies and new ideas by U S companies. Many of the institute's scientists and engineers focus on the fastest growing and potentially most profitable areas of science. These areas include advanced materials, electronics, superconductivity, automation, biotechnology, and thin layer technology.

The Materials Science and Engineering Laboratory (MSEL) at NIST investigates all classes of advanced materials. Some of these classes include ceramics, polymers, composites, and metallic alloys. The results are data, measurement tools, and services for understanding, improving, predicting, and controlling the processing and performance of materials.

THE INTERNATIONAL ORGANIZATION FOR STANDARDIZATION (ISO)

The International Organization for Standardization (ISO) establishes worldwide qualitative and quantitative standards for materials and processes. ISO is headquartered in Geneva, Switzerland, and consists of 91 national standards organizations from around the world. The United States representative to ISO is the American National Standards Institute. ISO has issued more than six thousand standards. These standards are proposed by 176 technical committees and 2,698 subcommittees.

SUMMARY

Materials have been used for millions of years to extend human potential. Before recorded history, humans used stones shaped by natural processes as weapons and tools to aid in survival. Eventually, humans learned to shape the naturally occurring materials found in their environment. The discovery of fire led to the use of new materials and the development of processes that further extended the human potential and improved the standard of living.

Today, students of material science and technology are given the opportunity to learn about the fascinating world of natural and synthetic materials used in all facets of society. The pace of change in this field is accelerating at an unbelievable rate. Today, materials can be engineered to meet nearly any operational condition.

At the same time, students must learn to be socially and environmentally responsible in making decisions about materials selected for all applications. Mistakes made in the past are ones we cannot afford to make in the future. We all must be more responsible in using and recycling the materials that attempt to improve standards of living.

IMPORTANT TERMS

Chemical Properties

Concurrent Engineering

Eoliths

Mechanical Properties

Nonbiodegradable

Nonrenewable Resources

Operational Conditions

Physical Properties

Rapid Prototyping

Reliability

Renewable Resources

Shelf Life

Specifications

Standards

Stereolithography

Time to Market

Trade Associations

QUESTIONS FOR REVIEW AND DISCUSSION

1. What materials were the earliest tools made from? How were they formed?
2. What contributions did the discovery of fire make to materials technology?
3. What four metals were the earliest used by humans? Why?
4. Explain the differences between renewable and nonrenewable resources. Give examples of each.
5. Briefly describe the advantages of recycling.
6. What social and economic factors must be considered when designing new products?
7. What is meant by concurrent engineering? How does it improve the process of developing new product lines?
8. What is rapid prototyping and how does it influence the design process?
9. What factors are generally considered in selecting materials?
10. What is the difference between a standard and a specification?
11. Who establishes standards? Who establishes specifications?

FURTHER READING

1. Earth Works Group. The Recycler's Handbook. Berkley, California: Earth Works Press (1990).
2. US Department of the Interior, Bureau of Mines. The New Materials Society, Vol. 3. Washington, D.C.: US Department of the Interior (1991).

INTERNET RESOURCES

http://www.astm.org
American Society for Testing and Materials

http://www.ansi.org
American National Standards Institute

http://www.msel.nist.gov
National Institute for Standards and Technology

http://www.iso.ch
International Organization for Standardization

http://www.em.doe.gov
U.S. Department of Energy

http://www.epa.gov
U.S. Environmental Protection Agency

2

Structure of Materials

KEY CONCEPTS

Upon completion of this chapter, you should understand:
➤ Basic atomic structure of matter.
➤ Chemical bonds and how they relate to properties.
➤ Methods of classifying matter.
➤ The periodic table of elements.

Why do certain materials conduct thermal and electrical energy while others insulate? Why are some materials more reactive than others? A basic understanding of the structure of matter is key to understanding why materials have the properties they do. The importance of understanding the basic atomic and molecular structure of materials cannot be overstated. All mechanical, physical, and chemical properties of materials are a direct result of their atomic structure. An understanding of the basic structure of each family of materials provides important information for engineers and technologists as they select materials for specific applications.

BUILDING BLOCKS OF MATTER

Approximately 110 chemical elements have been discovered in nature or synthesized in a laboratory. Elements cannot be subdivided into other substances. These 110 chemical elements comprise the gas, liquid, and solids that constitute the earth and its atmosphere as we know it. Some of these elements are far more plentiful than others. For example, two of the elements, oxygen (O) and silicon (Si) make up more than 75% of all solid matter. Oxygen, of course, is found in the air we breathe, but it is also in the ores from which metals are refined and in liquids we drink. Some elements are very rare and others only exist in the laboratory. Of the 110 chemical elements, eight constitute more than 97% of all earth substances. These eight include aluminum (Al), iron (Fe), calcium (Ca), sodium (Na), potassium (K), magnesium (Mg), oxygen (O), and silicon (Si).

Figure 2-1.
The Periodic Table of Elements provides basic information about elements.

Key to Chart:
- 26 — Atomic Number
- Fe — Element Symbol
- 55.847 — Atomic Weight

1 IA	2 IIA	3 IIIA	4 IVA	5 VA	6 VIA	7 VIIA	8 VIIIA	9 VIIIA	10 VIIIA	11 IB	12 IIB	13 IIIB	14 IVB	15 VB	16 VIB	17 VIIB	18 VIIIB
1 H 1.008																	2 He 4.003
3 Li 6.941	4 Be 9.0122											5 B 10.811	6 C 12.011	7 N 14.007	8 O 15.999	9 F 18.998	10 Ne 20.179
11 Na 22.989	12 Mg 24.305											13 Al 26.982	14 Si 28.086	15 P 30.974	16 S 32.066	17 Cl 35.453	18 Ar 39.948
19 K 39.098	20 Ca 40.078	21 Sc 44.956	22 Ti 47.88	23 V 50.942	24 Cr 51.996	25 Mn 54.938	26 Fe 55.847	27 Co 58.933	28 Ni 58.69	29 Cu 63.546	30 Zn 65.39	31 Ga 69.72	32 Ge 72.59	33 As 74.922	34 Se 78.96	35 Br 79.904	36 Kr 83.80
37 Rb 85.468	38 Sr 87.62	39 Y 88.906	40 Zr 91.224	41 Nb 92.906	42 Mo 95.94	43 Tc (98)	44 Ru 101.07	45 Rh 102.906	46 Pd 106.42	47 Ag 107.868	48 Cd 112.41	49 In 114.82	50 Sn 118.710	51 Sb 121.75	52 Te 127.60	53 I 126.905	54 Xe 131.29
55 Cs 132.905	56 Ba 137.33	57 La* 138.906	72 Hf 178.49	73 Ta 180.948	74 W 183.85	75 Re 186.207	76 Os 190.2	77 Ir 192.22	78 Pt 195.08	79 Au 196.967	80 Hg 200.59	81 Tl 204.383	82 Pb 207.2	83 Bi 208.980	84 Po (209)	85 At (210)	86 Rn (222)
87 Fr (223)	88 Ra 226.025	89 Ac† 227.028	104 Rf (261)	105 Db (262)	106 Sg (263)	107 Bh (262)	108 Hs (265)	109 Mt (266)	110 (269)	111 (272)	112 (277)						

* Lanthanide series:

58 Ce 140.12	59 Pr 140.908	60 Nd 144.24	61 Pm (145)	62 Sm 150.36	63 Eu 151.96	64 Gd 157.25	65 Tb 158.925	66 Dy 162.50	67 Ho 164.931	68 Er 167.26	69 Tm 168.934	70 Yb 173.04	71 Lu 174.967

† Actinide series:

90 Th 232.038	91 Pa 231.036	92 U 238.029	93 Np 237.048	94 Pu (244)	95 Am (243)	96 Cm (247)	97 Bk (247)	98 Cf (251)	99 Es (252)	100 Fm (257)	101 Md (258)	102 No (259)	103 Lr (260)

All elements are scientifically ordered and recorded in a table called the *Periodic Table of Elements,* or periodic table, Figure 2-1. The elements are arranged in ascending order in rows and columns according to their atomic number. The *atomic number* is the number of protons in the nucleus of an atom. Hydrogen (H) has an atomic number of 1. It is the first element in the table (upper-left corner). The next element is helium (He) with an atomic number of 2. An element's position in the table also provides information about the structure of the atom. Atomic structure is discussed in more detail in the next section.

The rows are known as *periods.* Periods are organized by number of electron shells. Electron shells are the orbits around the nucleus that contain the electrons. Period 1 is the top row in the table. All elements in Period 1 have one electron shell. The elements in Period 2 have two electron shells, and so forth.

The columns are known as *groups.* All elements in a group have the same number of valence electrons. Valence electrons are electrons in the outermost shell of an atom. Groups are labeled using Roman numerals I through VIII.

The periodic table provides additional information about each of the elements. The information reveals a lot about the properties and characteristics of the elements. For example, the density is an indication of the weight of an element. Gold (Au) has a density of 19.32 g/cc. Gold would obviously not be a choice where lightweight is a requirement. Melting point and boiling point are also given on some periodic tables.

ATOMIC STRUCTURE

All materials can be reduced to three basic building blocks—atoms, ions, and molecules. All matter is formed by the chemical bonding of these three building blocks of nature. Atoms are the basic building blocks of matter. They have a neutral electrical charge. Ions are similar to atoms, however, they have either a negative or positive electrical charge. Molecules are groups of atoms bonded together to form a material.

ATOMS

The atom is the smallest particle of an element that can retain the physical and chemical properties of the element. An atom consist of a nucleus and surrounding orbits which contain electrons, Figure 2-2. The nucleus is the densest part of the atom and consists of neutrons and protons.

Neutrons are subatomic particles with no electrical charge. *Protons* are positively charged subatomic particles. The number of protons and electrons in an atom are equal. This creates a neutral electrical charge for the atom.

As stated earlier, the atomic number of an element is the number of protons in its nucleus. The *atomic weight* of an atom is equal to the sum of the weight of the protons and neutrons in the nucleus of the atom. These values are given in the periodic table.

Figure 2-2.
The basic atomic structure of
an atom is made from protons,
neutrons, and electrons.

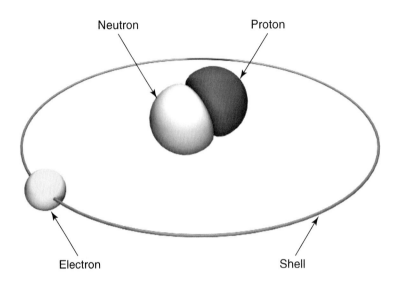

Figure 2-2.
The basic atomic structure of
an atom is made from protons,
neutrons, and electrons.

Atomic Mass

The *atomic mass* of an element is the average weight of all isotopes of that element. Isotopes are covered later in this chapter. Atomic mass is commonly indicated in unified atomic mass units (u). An *atomic mass unit (u)* is defined as 1/12 the mass of carbon 12. Carbon 12 has six protons and six neutrons and a mass of 12u.

The relative atomic mass of an element is Avogadro's number of atoms of the element. Atomic mass is measured in grams. *Avogadro's number* is 6.023×10^{23} and represents the number of atoms in one mole of the element. A *mole (mol)* is defined as having the mass in grams of the relative molar mass of that element. For example, one gram-mole of carbon 12 (^{12}C) has a mass of 12 grams on this scale. One gram-mole of sodium (Na) has a mass of 22.9898 grams. Both examples contain 6.023×10^{23} atoms. The atomic mass of an element is given in the periodic table.

Energy Levels

Surrounding the nucleus are orbits or shells. These are called *energy levels* and contain the electrons in the atom. The outermost orbit or shell is called the *valence shell.*

The energy levels are identified using capital letters beginning with K. The letter K designates the energy level closest to the nucleus, Figure 2-3. Each energy level is limited to a specific number of electrons. The K energy level can hold only two electrons at any one time, while the L energy level has a maximum of 8 electrons. The formula $2n^2$ is used to determine the maximum number of electrons for any of the energy levels. The variable n in the formula is the number of orbits from the nucleus. For example, to determine the maximum number of electrons for the M energy level, the value 3 (K, L, M) is substituted for n.

$$2n^2 = 2(3^2) = 2(9) = 18$$

Figure 2-3.
An atom can contain many
energy levels. This atom is
zinc.

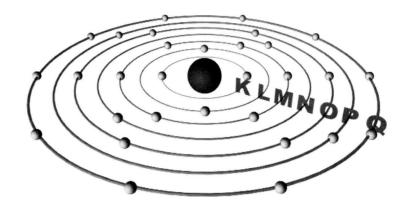

Figure 2-4.
This table shows the different
electron energy levels.

ELECTRON ENERGY LEVELS		
Energy Level	Energy Sublevel (Maximum Electrons)	Maximum Electrons
K	S(2)	2
L	S(2), P(6)	8
M	S(2), P(6), d(10)	18
N	S(2), P(6), d(10), f(14)	32
O	S(2), P(6), d(10), f(14), g(18)	50
P	S(2), P(6), d(10), f(14), g(18), h(22)	72
Q	S(2), P(6), d(10), f(14), g(18), h(22), i(26)	98

Energy levels should not be thought of as one dimensional. Electrons orbit the nucleus above or below one another in sublevels, which are also designated with letters. Figure 2-4 shows the levels and sublevels with the maximum number of electrons in each. Determining the electron configuration of an atom allows scientists to predict the reaction of one atom to another.

Isotopes

All atoms of an element have the same number of protons. However, the nucleus of some atoms of the element may have a different number of neutrons. These atoms are referred to as *isotopes.* The weight of an isotope differs from the basic element, however, the electrical charge is the same. For example, carbon 14 (^{14}C) is an isotope of carbon 12 (^{12}C).

IONS

An atom in a balanced state has the same number of electrons and protons. Therefore, it has a neutral electrical charge. When atoms are brought into contact with one another, they often give up or gain electrons in the outer shell (valence shell). When this occurs, the atom no longer has a neutral electric charge. This process is known as *ionization* and results in negative and positive ions. An *ion* is an atom with an unbalanced electrical charge.

An atom that loses one electron from its valence shell is called a *cation.* A cation has a positive electrical charge, since there is one more proton than electron. An ion that acquires additional electrons is called an *anion.* An anion has a negative electrical charge, since it has more electrons than protons.

MOLECULES

Molecules are groups of atoms or ions bonded together to form a material. A common molecular substance is water. Most people know water is made from molecules of hydrogen (H) and oxygen (O) and its chemical designation is H_2O. This means that for each oxygen atom there are two hydrogen atoms.

Molecules vary in size. One molecule of water contains only three atoms. However, one molecule of polyethylene ($C_{100}H_{202}$) has 100 atoms of carbon (C) and 202 atoms of hydrogen (H).

Some material families rely on molecules as their basic building blocks. In other words, the molecule, not the atom, determines the properties of the material. Polymers belong to a family of materials based on molecules. For this reason, polymers are sometimes referred to as *molecular materials.*

CHEMICAL BONDS

Properties and characteristics that make materials different can often be explained by the type of bond that attracts atoms and molecules to one another. The bonds between atoms, ions, or molecules form between the outer orbit or valence shells of atoms. At the most basic level, matter wants to stabilize or reach equilibrium. When the valence shell of an atom is not full, it attempts to fill its orbit by attracting electrons from other atoms. An atom may "give up" an electron if its valence shell only has a few electrons. Since atoms do not actually give up or take on electrons, the electrons are shared between atoms. This sharing creates a chemical bond holding the atoms together.

A filled outer energy shell is the state of lowest energy. This condition, therefore, produces materials that are stable and resist change. Some elements have atoms with complete or filled outer energy shells. Those elements are very stable and do not react with other elements. The term used to describe elements with filled valence shells is *inert.* Helium (He) and argon (Ar) are examples of inert elements. Helium has one energy shell containing two electrons. This is the maximum number of electrons allowed in shell K, Figure 2-5.

The types of bonding are generally classified as ionic, covalent, metallic, and secondary. The first three are considered types of primary bonding. *Primary bonds* are atom-to-atom bonds. *Secondary bonding,* or molecule-to-molecule bonding, includes van der Waals forces.

Figure 2-5.
A helium atom has one energy
shell containing two electrons.

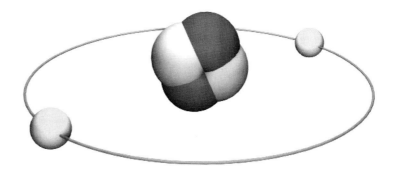

IONIC BONDING

Ionic bonding takes place between atoms that have few electrons in their valence shell and those that have nearly complete shells. When these atoms come close to each other, both have a tendency to react in order to reach the lowest energy level. The atom with few electrons is willing to "give" its valence electrons to the atom whose valence shell is nearly complete. This creates a stable condition as the outer orbits of both atoms are then filled by the shared electron.

Sodium chloride (NaCl) provides an excellent example of ionic bonding, Figure 2-6. Sodium (Na) has only one electron in its valence shell. It "gives" the single valence electron to chlorine (Cl), which has seven valence electrons. In this process, sodium becomes a positive ion and chlorine becomes a negative ion. A tight bond is created because of the attraction between oppositely charged ions.

Ceramic materials tend to be bonded ionically. Ceramics are usually combinations of metal and nonmetal elements. Metals usually have few electrons in their valence shells and nonmetal elements nearly full valence shells. This ionic bond is a reason for the high melting temperatures, hardness, and low thermal and electrical conductivity of ceramics.

Figure 2-6.
Sodium chloride is held together by ionic bonding. One electron is shared between the two atoms.

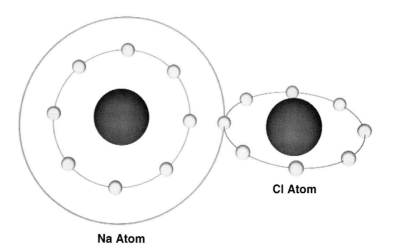

Cl Atom

Na Atom

COVALENT BONDING

Elements that bond covalently are those whose valence shell is about half full. These elements include carbon (C), nitrogen (N), and others found in groups IVA and VA of the periodic table. These atoms do not want to give up electrons, yet cannot strongly attract electrons. Therefore, they share multiple electrons in order to try to reach the lowest energy level.

Diamond is a good example of a material that is bonded covalently. Diamond is a carbon-based, naturally occurring material. It is the hardest natural material known to exist. It has an extremely high melting temperature and is very stable. Figure 2-7 illustrates the covalent bond formed when carbon atoms combine.

METALLIC BONDING

Metal elements tend to have atoms whose valence shells have only a few electrons. When a large number of metallic atoms are combined, the valence electrons from individual atoms form an "electron cloud." This cloud moves around in the material trying to make each atom reach equilibrium at its lowest energy level.

This movement of the electron cloud is responsible for many of the characteristics of metal. Unlike ceramics, metals are good conductors of electrical and thermal energy.

SECONDARY BONDS

As mentioned previously, the building blocks for certain materials are molecules. For example, the structure of polymers is based on the molecule. In the case of most plastics, groups of large molecules are bound together to form long, chainlike patterns. The combining of one molecule to another is accomplished through secondary bonding. *Secondary bonds* form through weak attractions of positive and negative charges of molecules. The attraction that results in secondary bonds is called *van der Waals forces.* Secondary bonds are found in plastics, woods, and other molecular materials.

As a result of these weak bonds, molecular materials have low strength and low melting temperatures, when compared to metals and ceramics. These materials are also poor conductors of heat and electricity.

Figure 2-7.
These atoms of carbon are held together by covalent bonding. Two electrons are shared between the atoms.

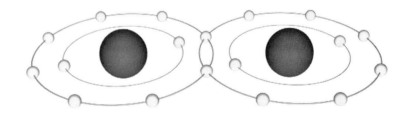

SOLID MATERIAL STRUCTURE

This book focuses primarily on solid materials (metal, ceramics, wood, etc.) rather than liquids and gases. Solids have structures obviously different from gases and liquids. These structures are the framework of solids and provide them with their own, individual characteristics. Solids exist in one of two forms: crystalline or amorphous.

Crystalline solids have ordered, three-dimensional geometric patterns that repeat throughout the material. Crystalline structures are found in most material classifications.

An *amorphous solid* does not have the repeating, three-dimensional geometric patterns. By definition, amorphous means having no definite form. The classic example of an amorphous solid is glass. The structure of glass is without order and has a nonrepeating pattern.

UNIT CELLS

The repeating patterns of three-dimensional structures in crystalline solids are formed by what are termed *unit cells.* The geometry of the unit cells may be one of fourteen types, Figure 2-8. Of these fourteen, only three or four are common. These common types include body centered cubic (BCC), face centered cubic (FCC), close packed hexagonal (HCP), and the simple tetragonal unit cell. Unit cells form a three-dimensional configuration called a *lattice.*

Figure 2-8.
A unit cell in the atomic structure of solid matter can take one of many shapes. Six of the possible shapes are shown here. The most common shapes are BCC, FCC, and HCP.

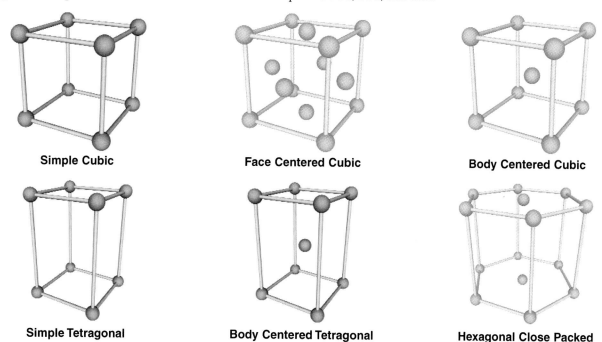

Simple Cubic   **Face Centered Cubic**   **Body Centered Cubic**

Simple Tetragonal   **Body Centered Tetragonal**   **Hexagonal Close Packed**

A single material may exist in more than one of these configurations under certain conditions. For example, carbon steel has a BCC structure at ambient temperature. At approximately 1400° F, it makes a transformation to a FCC structure. At approximately 2500° F, it transforms back to a BCC structure. If allowed to cool naturally, it will reverse these transformations at each of these temperatures. Materials that can make this type of transformation in both directions, such as carbon steel, are called *allotropic.* Materials that can make these transformations in one direction only are known as *polymorphs.* Allotropic materials are also polymorphs.

CLASSIFYING MATERIALS

There are virtually thousands of materials with different properties and characteristics. Grouping or classifying these materials by common characteristics assists designers, scientists, and engineers when selecting materials for specific applications.

Several systems for classifying materials have been devised. The most basic perhaps, is to classify materials as metals and nonmetals. Although not incorrect, this is probably too simple for most purposes. A broader system is often used to classify all matter as liquid, solid, and gases. This system is certainly accurate, but for the purposes of this book, it is still too simple. In addition, this book focuses on solids used for industry. Therefore, this book classifies solid materials as:

- Metals
- Polymers
- Ceramics
- Composites

Although this system may not include every solid material in existence, it provides a method of organizing solid materials for systematic study. Each of these material families can be subclassified by similarities of properties and characteristics. Figure 2-9 presents each family of solid material with subclassifications used in this book.

METALS

Metals have been used to extend human potential for thousands of years. Metals found free in nature, such as copper, gold, silver, and meteoric iron, were the earliest metals used. Originally, they were used for ornamentation and decoration. Eventually, however, humans discovered that metals could be cut and cast to form shapes for many purposes. The significance of metals in human history can easily be demonstrated through the names given to early human development, such as the Bronze Age and the Iron Age.

Today, evidence of the importance of metals in society can be found all around us. Metals are in the homes we live in, the cars we drive, and the schools we attend. Although the raw materials for metals are not as plentiful as they once were, the demand is still as high as ever. Recycling helps reduce the need for metal from raw ore.

Metals, as with most industrial materials, are defined by their properties. Metals as a material family are hard, tough, and strong. They also have high electrical and heat conductance. Metals are generally classified as ferrous or nonferrous.

Figure 2-9.
This table shows the
classification of various
industrial materials.

INDUSTRIAL MATERIAL CLASSIFICATION SYSTEM		
Material Family	**Subclassification**	**Examples**
Metals	Ferrous	Iron Steel
	Nonferrous	Magnesium Aluminum Copper Copper Alloys Zinc Noble Metals Gold Silver
Polymers	Synthetic	Plastics Adhesives Elastomers
	Natural	Wood Rubber Paper
Ceramics	Natural	Flint Emery Clay Garnet Diamond
	Synthetic	Aluminum Oxide Silicon Carbide Zirconia
	Glass	Soda-Lime Borosilicate
Composites	Polymer Matrix	Fiberglass PEEK Aramid Epoxy Plywood
	Metal Matrix	Graphite-Aluminum Boron-Aluminum Silicon-Carbide Titanium
	Ceramic Matrix	Concrete Lithium-Alumina-Silicate Magnesia-Alumina-Silicate

Ferrous metals are those that have significant iron content. This is defined as more than 50%. All ferrous metals attract magnetic materials. Ferrous metals include wrought iron, steel, and stainless steel.

Nonferrous metals have less than 50% iron content. They do not attract magnetic materials. Nonferrous metals have higher corrosion resistance than ferrous metals. Nonferrous metals include aluminum, copper, copper alloys, zinc, lead, titanium, tin, and magnesium.

POLYMERS

The term *polymer* is from the Greek *poly* meaning many and *mer* meaning parts. Polymers are molecular-based materials formed from carbon atoms. Long, chainlike patterns of molecules are the foundation for polymeric materials. The terms polymer and plastic are often used interchangeably, but this is incorrect. Polymer is a term used to describe a broad family of materials that includes plastic.

The term *plastic* refers to only synthetic polymers. Plastics are further classified as thermosetting plastics (or thermosets) and thermoplastics. The basic difference is in the ability to be remelted and reprocessed. *Thermosets* bond in such a way that does not permit them to be remelted after initial processing. *Thermoplastics,* on the other hand, can be remelted and reprocessed.

Other polymers include natural materials. Wood, rubber, and leather are examples of natural polymers. These materials are molecular materials, like plastics. As such, they exhibit many of the same properties as plastics.

Polymers as a family do not have high strength. However, polymers do have high resistance to electrical conductivity, and low melting and service temperatures.

CERAMICS

The word ceramic is derived from the Greek word *keramos.* This translates to burned material. Ceramic materials are brittle. They are usually inert and nonconducting materials. They have a wide range of uses in the arts, construction, and manufacturing.

Ceramics have crystalline and noncrystalline structures. Glass, as stated previously, is the classic example of a noncrystalline ceramic. Ceramics include cement and concrete, as well as the stoneware used for dinner plates and the porcelain fixtures in your bathroom.

Advanced ceramics are now being used in the aerospace and automotive industries. The tile applied to the underside of the space shuttle is pure silicon ceramic that can resist re-entry temperatures of 2200° F. The automotive industry is now developing engines with ceramic lined cylinder walls to improve efficiency and reduce wear.

The future for ceramics is a bright one. Researchers are providing new applications for ceramics regularly. Their high service temperatures, low conductivity, and hardness make them ideal for many applications.

COMPOSITES

The newest classification of materials is composites. *Composites* are two or more constituent materials combined in order to provide a mixed material with useful properties for specific applications. Each constituent material, or "ingredient," has its own, usually unique, characteristics.

Composites consist of a *reinforcing material* and a *binder.* For example, fiberglass is a composite of glass fibers (reinforcement) in polyester or other plastic resin (binder). Composites are generally classified as polymer matrix composites (PMC), metal matrix composites (MMC), or ceramic matrix composites (CMC).

SUMMARY

All matter, be it liquid, gas, or solid, is made up of one of three basic building blocks: atoms, ions, or molecules. Atoms have neutral charges and consist of subatomic particles called protons, neutrons, and electrons. Neutrons and protons reside in the center core of the atom, known as the nucleus. The nucleus has the greatest mass of the atom. Surrounding the nucleus are orbits called energy levels or electron shells. Electrons move around the nucleus in these orbits.

The attraction between atoms, ions, or molecules of different elements is important in defining material properties. Most ceramics and metals have strong chemical bonds and, therefore, are strong and have higher melting temperatures. Molecular materials, such as plastic and wood, are relatively weaker materials with lower service temperatures.

Solid materials are either crystalline or noncrystalline. Crystalline solids have geometric structures that repeat themselves throughout the material. These structures are made of unit cells that form a lattice of three-dimensional shape. These lattice structures form individual crystals of solid matter.

Classifying solid materials as polymer, metal, ceramic, or composite is a way of organizing or grouping materials by properties and characteristics. These classifications can then be further classified for systematic study.

IMPORTANT TERMS

Allotropic	Isotopes
Amorphous Solid	Lattice
Anion	Mole (mol)
Atomic Mass	Molecular Materials
Atomic Mass Unit (u)	Neutrons
Atomic Number	Nonferrous Metals
Atomic Weight	Periodic Table of Elements
Avogadro's Number	Periods
Binder	Plastic
Cation	Polymer
Ceramics	Polymorphs
Composites	Protons
Crystalline Solids	Reinforcing Material
Energy Levels	Secondary Bonds
Ferrous Metals	Thermoplastics
Groups	Thermosets
Inert	Unit Cells
Ion	Valence Shell
Ionization	van der Waals Forces

QUESTIONS FOR REVIEW AND DISCUSSION

1. How is the atomic weight of an atom determined?
2. Describe the arrangement of chemical elements in the period table. What is the difference between a period and a group?
3. What is the difference between an atom and an ion?
4. How are molecules different than ions or atoms?
5. What is an isotope?
6. What are the basic components of an atom?
7. What does inert mean?
8. How many electrons can electron shell P hold? Use the formula $2n^2$ to answer the question.
9. Describe ionic bonding. Give an example of materials based on ionic bonding.
10. In which chemical bonding method is an "electron cloud" formed?
11. Explain the difference between a cation and anion.
12. What does the term amorphous mean? Give an example of an amorphous material.
13. List three common unit cell geometric shapes.
14. To what does the term polymorph or polymorphic refer?
15. Define allotropic.

FURTHER READINGS

1. Brady, George S. and Clauser, Henry R. Materials Handbook (13th edition). New York: McGraw-Hill Inc. (1991).
2. Oskeland, Donald R. The Science and Engineering of Materials (3rd edition). Boston: PWS Publishing Co. (1994).

INTERNET RESOURCES

http://www.particleadventure.org
The Particle Data Group of the Lawrence Berkeley National Laboratory

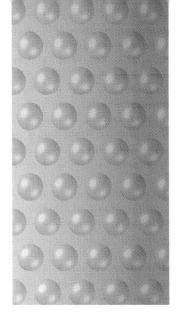

3

Mechanical Properties of Materials

KEY CONCEPTS

Upon completion of this chapter, you should understand:
➢ Stress and strain.
➢ Elastic and plastic deformation of materials.
➢ Material strengths, such as tensile, compressive, and shear strengths.
➢ Ductility and fracture toughness of materials.
➢ Basic product design criteria.

To select a material for any application, it is essential to understand the properties of the material. This is required to ensure the functionality and safety of the product. A material should be strong enough to support the expected load. In other words, the selected material must have strength higher than the expected stress on the product. It is impossible to design any safe product without knowing the material properties.

The most important material property is the mechanical property. *Mechanical property* is the response of a material to mechanical forces or stresses. Mechanical property usually includes strength, hardness, elasticity, plasticity, ductility, fracture toughness, and so on. This chapter introduces fundamentals and applications of mechanical properties of materials.

STRESS AND STRAIN

A material's response to external or internal force is measured by stress. The intensity of force normal to the surface of a material is defined as *normal stress.* Normal stress is commonly denoted by Greek letter σ (sigma). Figure 3-1 illustrates a column with a cross sectional area of A_o, subjected to a normal force P.

The normal stress in the column, σ, is defined as:

$$\sigma = \frac{P}{A_o}$$

where the stress σ is measured in pound per square inch (psi) in the US Customary system or Newton per square meter (N/m^2, which is also known as Pascal or Pa). For example, if a column has a cross sectional area

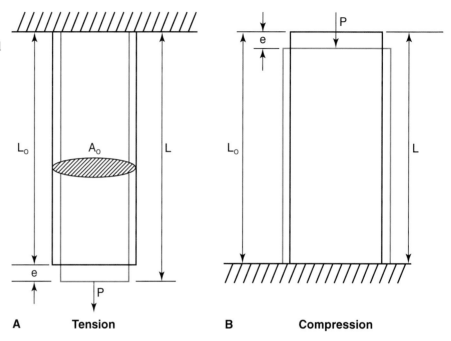

Figure 3-1.
Normal stress in tension (A) and compression (B) is defined as the applied load divided by the original cross sectional area of the column.

(A_o) of 2 in^2 and the column is loaded (P) with 500 lb., the normal stress (σ) is 250 psi. A normal stress can be either tension, Figure 3-2A, or compression, Figure 3-2B.

A stress on a material causes the material to deform. The deformation on the material is represented by strain. As in Figure 3-1A, the column was stretched by a normal force in the amount of "e." The material's elongation under load is:

$$e = L - L_o$$

where L and L_o are final and original lengths of the column, respectively. For example, if a column with original length (L_o) of 2″ is stretched to 2.005″, the elongation is 0.005″.

The *strain* on the column, ε (epsilon), is defined as the deformation per unit length:

$$\varepsilon = \frac{L - L_o}{L_o}$$

Strain is a dimensionless value, although in practice, units such as in/in or mm/mm are often used. For example, if a column with original length (L_o) of 2″ is stretched to 2.005″, the strain on the column (ε) is 0.0025.

SHEAR STRESS AND STRAIN

Besides normal stress, there is another type of stress known as shear stress. **Shear stress** is the intensity of the force parallel to the surface of a material. It is represented by the Greek letter τ (tau), Figure 3-2. For example:

$$\tau = \frac{P}{A_o}$$

Figure 3-2.
A bolt in the joint is subjected
to a direct shear stress.

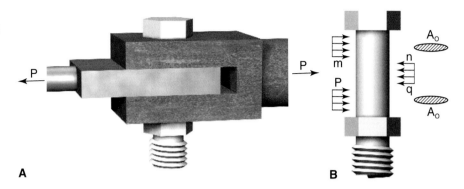

Figure 3-2.
A bolt in the joint is subjected
to a direct shear stress.

where P is the shear force and A_o is the cross sectional area of the shearing plane. If a bolt with a cross sectional area (A_o) of 0.2 in² is sheared with a force (P) of 100 lb., the shear stress (τ) is 500 psi.

Shear stress causes distortion of the element it acts upon, as shown in Figure 3-3. **Shear strain** is defined as the distance (e) along the shear direction divided by the distance (L) perpendicular to the shear direction. Numerically, the shear strain is equal to the angular change of the element, which is represented by lower-case Greek letter γ (gamma).

$$\gamma = \frac{e}{L}$$

For example, if an element 2" long (L) is distorted for 0.002" (e) by a shearing force, the shear strain is 0.001.

MODULUS OF ELASTICITY

The stress-strain behavior of an industrial material is representative of the mechanical properties of the material. It is determined by tests performed on machined specimens of the material. A typical tensile test machine is shown in Figure 3-4. The test specimen is installed between the two grips of the testing machine. The testing machine can measure the load applied on the specimen. An **extensometer** is a sensor that measures elongation of the specimen. It can be mounted on the middle section of the specimen.

Figure 3-3.
Shear strain is measured as the ratio of the distance "e" along the shear direction to the distance "L" perpendicular to the shear direction.

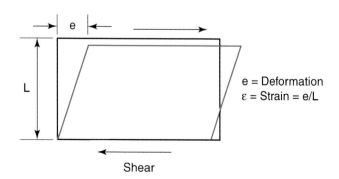

e = Deformation
ε = Strain = e/L

Shear

Figure 3-4.
A typical tensile test machine can measure and record the load and deformation of the tested specimen. (MTS)

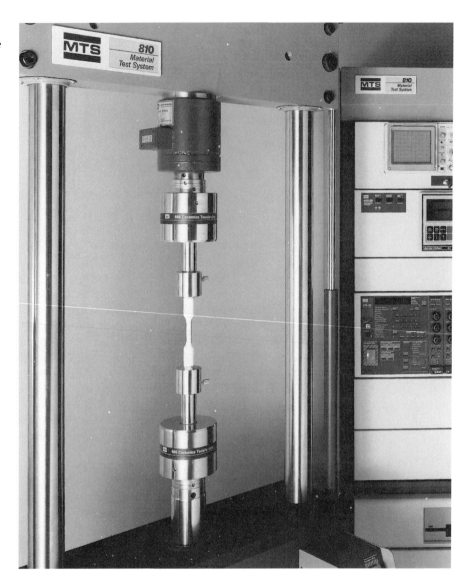

Figure 3-5 shows a typical stress-strain curve of a structural steel when a load is slowly applied onto the material. Strains are plotted on the horizontal axis and stresses on the vertical axis. This graph also illustrates what happens physically on the test specimen as the specimen is stretched. The diagram begins with a straight line from O (origin) to E, which means that the stress and strain are proportional. This region is known as elastic range, with the maximum stress at E being called elastic limit. The proportional relationship is known as Hooke's law, which is expressed mathematically as:

$\sigma = E\varepsilon$

where σ is stress and ε is strain. E is a proportional constant. This constant is known as *modulus of elasticity.*

Figure 3-5.
This typical stress-strain curve for a structural steel shows representations of the related dimensional change in the test specimen.

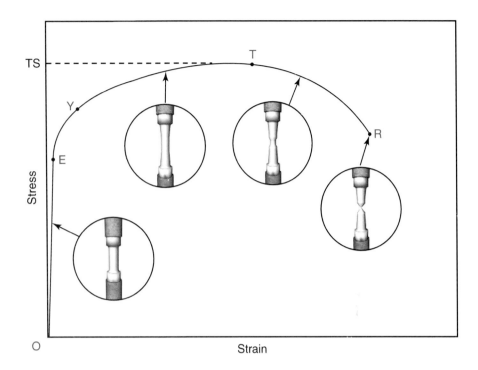

Modulus of elasticity can be obtained by the slope of the straight line from O to E. It represents the rigidity of a material. A rigid material has low deformation under stress. The higher the modulus of elasticity, the more rigid the material is. The modulus of elasticity of AISI 1025 steel is 29,000,000 psi, whereas that of aluminum alloys is about 8,000,000 psi.

With an increase in the load beyond the elastic limit (E), the strain or deformation begins to increase much more rapidly than before as the stress increases. Starting at point Y on the stress-strain curve in Figure 3-5, considerable elongation occurs without noticeable increase in the tensile force. This phenomenon is known as the *yield* of the material. The point Y is called the *yield point.* The corresponding stress is known as yield stress or *yield strength* of the material.

After undergoing the large strains or deformation during yielding, the steel begins to strain harden. *Strain hardening* is when the material becomes stronger or harder due to the deformation. The material undergoes changes in its crystalline structure. This results in increased resistance of the material to further deformation. Additional elongation requires an increase in the tensile load up to the maximum value at point T in Figure 3-5. The maximum stress is called the ultimate stress or *ultimate tensile strength* of the material. After the maximum stress is reached, the cross section of the specimen drastically decreases. This is known as *necking.* The cross section continues to reduce with increased loading. Further stretching the bar is accompanied by a reduction in load, which leads to rupture of the specimen at point R, as shown in Figure 3-5.

ELASTICITY AND PLASTICITY

Using a typical tensile test, the modulus of elasticity, yield strength, ultimate tensile strength and rupture strength of the material can be determined. The deformation behavior of the material during the static loading process is important for industrial applications. Figure 3-6 shows the stress-strain diagrams of elastic and plastic deformation of materials. When a load is applied to the specimen, the stress and strain increase from O to E. If the load is released from any point between O and E, the material follows the same route back to the origin O, as shown in Figure 3-6A. There is no retained deformation or residual strain on the material. Elasticity is the ability of a material to return to its original dimensions after unloading. The material itself is said to be *elastic.* If the stressed material remains in the elastic range, it can be loaded and unloaded repeatedly without significantly changing its properties.

However, if the material is loaded to exceed the yield point, point B for example, upon unloading, the material follows line BC on the diagram, Figure 3-6B. This unloading line is typically parallel to the initial straight line (OE) of the stress-strain curve. When the load is completely removed and point C is reached, a *residual strain* or *permanent strain* measured by line OC remains in the material. If the material is loaded again, the loading begins at C on the diagram and continues to B. This is the point at which unloading began during the first loading cycle. The material then follows the original stress-strain curve toward points T and R. The characteristic of a material by which it undergoes inelastic strains beyond the stress of the elastic limit (E) is known as *plasticity.* When large deformation occurs in a ductile material loaded into the plastic region, the material is said to have undergone *plastic deformation.*

Most industrial machines are designed so that the materials will work in the elastic range. In this way, there is no permanent deformation on the machine after assembly and the precision of the machine can be maintained

Figure 3-6.
These stress-strain diagrams illustrate elastic behavior (A) and plastic behavior (B).

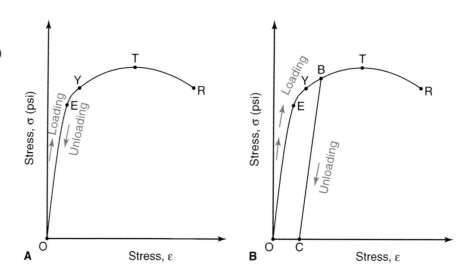

as expected. Since the elastic limit and yield strength are very close, and yield strength is easily determined, yield strength is commonly used for designing a product. Materials are in the plastic region during the manufacturing process, for example, rolling or extrusion. The plastic deformation process causes work hardening in most of metals and therefore increases the strength of the material. That is why cold rolled steel is stronger than hot rolled steel with the same composition.

TENSILE STRENGTH

During a tension test, the maximum stress reached (point T in Figure 3-5) before breaking is called the *ultimate tensile strength* of the material. It is calculated by dividing the maximum load of the material by its original cross-sectional area. Most metals have noticeable plastic deformation before they reach the maximum stress. However, brittle materials, like most ceramics, have very limited plastic deformation and rupture soon after exceeding the elastic range.

Polymers have a wide range of mechanical properties. Some plastics are hard and brittle. These reach the ultimate tensile strength soon after exceeding elastic limits, just like ceramics. Other plastics are very ductile. These can be stretched to several times their original length before breaking, like plastic garbage bags.

Tensile strength is usually used to evaluate the strength of a material because it is related to the atomic bond inside the material. It is also used for design, especially when it is difficult to decide the yield strength of the material, such as a ceramic. However, tensile strength is not used as frequently as yield strength. A disadvantage of using tensile strength as a design criterion is that materials are often useless when the tensile strength is reached.

COMPRESSIVE STRENGTH

Some materials, such as cast iron, have low tensile strength but can withstand high compressive stress. A machine base or building foundation are examples of applications where a material is subjected to high compressive stress. For these applications, it is necessary to understand the compressive behavior of the material. A compression test for materials is merely the opposite of the tension test, with respect to the direction of the applied load. The test is performed by applying a compressive load on the specimen until it fractures. During the test, the maximum compressive stress, known as *compression strength,* can be found.

In practice, for materials subjected to compression load, we must consider a factor known as buckling. If a column is compressed, it may bend and deflect laterally instead of failing by direct compression. This phenomenon is called *buckling.* Any material subject to compressive stress must possess the structure stability so that it will not fail by buckling. A proper ratio of cross sectional area to the length of the column must be chosen to ensure the structural stability.

SHEAR AND TORSION

A stress-strain diagram can be plotted with shear stress and shear strain. This stress-strain curve is very similar to the curve of normal loading. As shown in Figure 3-7, there is a straight line or elastic range followed by plastic deformation. In the elastic range, there is a linear relationship between the shear stress and shear strain. That is, Hooke's law is followed for shear as in the case of tension:

$$\tau = G\gamma$$

where τ is shear stress and γ the shear strain. G is a proportional constant or the slope of the straight line, which is known as the **shear modulus of elasticity.** Generally, a tensile modulus of elasticity is 2.5 to 3 times the shear modulus of elasticity. For example, the tensile modulus of elasticity of AISI 1025 steel is 29,000,000 psi. The shear modulus of elasticity of the same material is 11,000,000 psi.

Yield and rupture strengths of shear can be determined by a shear test. Since tensile testing is much more commonly used and easier to perform than shear testing, it is often useful to approximate the shear values. For example, shear yield strength generally falls between 0.5 and 0.6 of the tensile yield strength.

Torsion refers to a twisting load placed on a structural member. Gear shafts and axle shafts in automobiles are two examples of components that are subjected to torsion. Torsion generates shear stress inside the material. The maximum shear stress a material can withstand before failure is known as **shear strength.** The amount of power or torque a shaft can transmit depends on the cross section of the shaft and the shear strength of the material.

DUCTILITY AND FRACTURE TOUGHNESS

A material that can undergo a large amount of plastic deformation before fracture is said to be **ductile.** The fracture is called a **ductile fracture.** On the other hand, a brittle material undergoes very limited plastic deformation before breaking. This fracture is called a **brittle fracture.**

Figure 3-7.
A shear stress-strain curve is similar to that of tension.

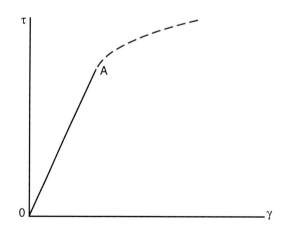

Normally, a brittle fracture occurs suddenly and often leads to a disaster. A ductile material is more resistant to impact or other dynamic loading than a brittle material. Therefore, under a dynamic loading condition, a material must have desirable ductility. For example, a white cast iron is hard and brittle. Therefore, it is not suitable for dynamic loading such as those found on shafts. A medium carbon steel is much more ductile and, therefore, more suited for a shaft.

The ductility of a material in tension can be characterized by its elongation and by the reduction in area of the cross section where fracture occurred. *Percent elongation* measures the total amount of elongation before fracture, relative to its original length. This is defined as:

$$\text{Percent Elongation} = \frac{L_f - L_o}{L_o} \times 100\%$$

where L_o and L_f are the original and final lengths of the test specimen, respectively. For example, if a material with original length of 2″ (L_o) is stretched to 2.30″ (L_f) when it fractures, the percent elongation is 15%. A material that has 5 percent or less elongation is considered a brittle material.

A steel may have an elongation about 10 to 40 percent, depending on its composition. Structural steel usually falls between 25 and 30 percent. The elongation in aluminum alloys varies from 1 to 45 percent, depending on the composition and treatment.

Another measurement of material ductility is *percent reduction in area.* This evaluates the change in the cross-sectional area of a specimen. It is defined as:

$$\text{Percent Reduction in Area} = \frac{A_o - A_f}{A_o} \times 100\%$$

where A_o and A_f are original and final cross sectional areas respectively at the fracture section of the test specimen. For example, if a material sample has an original cross sectional area of 0.2 in^2 and the cross section is reduced to 0.15 in^2 after fracture, the percent reduction in area is 25%. For ductile steels, the percent reduction in area can be as much as 50%. However, the percent reduction in area for cast iron usually cannot be measured.

There are always some cracks, flaws, or inclusions inside any material. A material will fracture if the crack spreads due to the stress on the material. As a result, the material will fail below its yield strength, especially for the materials of high strength. *Fracture toughness (K_{IC})* measures the resistance of a material to fracture. In other words, it is a measure of how well a material resists spreading, or propagation, of cracks. For a thick material with a small internal crack, the fracture toughness K_{IC} is defined as:

$$K_{IC} = \sigma_c \sqrt{\pi a}$$

where σ_c is the critical stress in the material as the cracking is started and "a" is the length of the existing crack inside the material. Fracture toughness (K_{IC}) has a unit of MN m$^{-3/2}$. For most metals and alloys, the fracture toughness K_{IC} is in the range from 20 to 150 MN m$^{-3/2}$. For alumina ceramics, K_{IC} is about 4.9 MN m$^{-3/2}$.

Fracture toughness is useful in material selection to prevent mechanical failure or fracture. This is especially true for materials having high yield and tensile strength, such as those used in aerospace applications. The larger the K_{IC}, the tougher the material is. In other words, the crack in the material with higher K_{IC} is less likely to propagate and cause a fracture. Values of K_{IC} are recorded in handbooks and material selection tables. In some severe applications, a required K_{IC} value for the material may be specified.

FATIGUE

The loads acting on a material greatly determine how a material behaves. In some situations, loads are gradually applied over a long period and change slowly. Such loads are known as *static loads.* While designing a structure to resist static loads, yield strength of the material is typically used. A safety factor is also incorporated to make certain that the material will not fail at the expected stress. The safety factor for metals is normally 1.5 to 2.0.

In some situations, components are subjected to dynamic loading. For example, a load is applied on a gear tooth when engaged with a matching gear. The load is released when the teeth are no longer in contact. Figure 3-8 shows some typical dynamic loading patterns occurring on mechanical components.

The loads in rotating machinery, automobiles, ships, and airplane structures may go through millions of cycles. This is called *cyclic loading.* A structure subjected to cyclic loading is likely to fail at a lower stress than when subjected to static loading.

Failure due to dynamic loads is called fatigue. *Fatigue* is the behavior of a material under cycles of stress and strain, causing a deterioration of the material that results in progressive cracking and eventual failure.

Figure 3-8.
Typical dynamic loading patterns. A—Repeated. B—Reversed. C—Fluctuating.

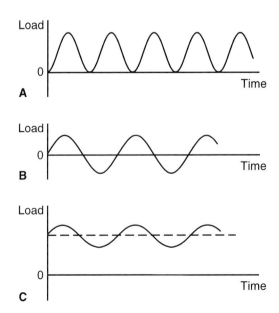

An example of fatigue failure is breaking a wire paper clip by repeatedly bending it back and forth. If bent only once, the wire does not break. If the load is reversed by bending the clip in the opposite direction, it still may not break. However, if the entire loading cycle is repeated several times, the clip will finally break.

The magnitude of the load or dynamic stress acting on a structure when a fatigue occurs is less than the magnitude of a static load. In order to determine the strength of a material under repeated loading, the material is tested at various cyclic stress levels. The loading cycle at failure is recorded at certain stress level. When a large enough group of data is gathered, a curve of stress versus number of cycles to failure can be plotted. This curve is called an *endurance curve,* also known as an *S-N diagram,* Figure 3-9. When the vertical coordinate is stress in a linear scale and the horizontal axis is number of cycles to failure in a logarithm scale.

An endurance curve shows that the smaller the stress, the larger the number of cycles for the material to fail. For some materials, like steel, the diagram has a horizontal asymptote known as the *fatigue limit* or *endurance limit.* Below this stress level, a fatigue failure will not occur regardless how many times the load is repeated. Thus, the fatigue limit is used as a design criterion for structural components subjected to cyclic loading. For other materials, such as aluminum, there are no distinct endurance limits. Failure stress continuously decreases with increasing number of loading cycles. In this case, an artificial fatigue limit is defined with certain loading cycles. For example, a fatigue limit of aluminum is usually taken as the failure stress at 5×10^8 cycles.

Since fatigue failures usually begin with a microscopic surface crack at a point of localized stress, the condition of the material surface is extremely important. A well-polished surface gives a higher fatigue limit for the material than an unpolished surface. In some industrial components, surface treatment, such as steel carbonizing, results in compressive stress on the surface. This can be beneficial in improving fatigue resistance. On the other hand, corrosive environment leads to deterioration in the fatigue limit of the material.

Figure 3-9.
A typical endurance curve, or S-N diagram, shows a fatigue limit of a material.

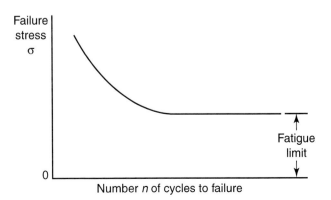

HARDNESS

Hardness is a measure of a material's resistance to surface plastic deformation by penetration or scratching. It is one of the most widely used factors in material selection and property control. Compared with other mechanical tests like tension test, hardness test can be conducted much faster. Similar materials may be graded according to hardness. Certain hardness values may be specified for certain service conditions. The quality level of materials or products can be checked and controlled by hardness testing. This is extremely useful in the quality control of steel heat treatment. Moreover, there is certain relationship between the hardness of a material and its other mechanical properties. By establishing a correlation between the hardness and tensile strength, a simple hardness test can serve to estimate or control the uniformity of tensile strength in the material.

There are many hardness tests. One of the best known is *Brinell hardness test.* This test consists of pressing a steel ball (usually 10 mm diameter) into the test piece and converting the measured impression area to a *Brinell Hardness Number (BHN).* A load of 3000 kg is used for hard metals, 1500 kg for metals of intermediate hardness, and between 500 kg and 100 kg for soft materials. The units of load and steel ball diameter are always recorded in metric units. The harder the material, the higher the Brinell hardness number.

Another common hardness test is the *Rockwell hardness test.* The hardness number is determined by the impression an indenter makes under a static load. However, the loads in a Rockwell hardness test are usually 60 kg, 100 kg, or 150 kg. The Rockwell hardness number is converted from the impression depth and can be read directly from a dial indicator or digital readout. Thus, it takes much less time to perform a Rockwell hardness test than Brinell hardness test.

In order to test the hardness of sheet material or close to the surface of a material, a *Rockwell superficial hardness test* is conducted. This test uses light loads of 15 kg, 30 kg, or 45 kg.

A more precise hardness test uses a device called a *microhardness tester.* This device has a precise diamond indenter that can be used to measure the hardness of microscopic particles or phases. The load used is very light, usually ranging from 10 g to 1,000 g. The process of microhardness test is similar to Brinell hardness test. The microhardness test is suitable for polymers, metals and ceramics.

A common hardness test for plastics and elastomers is *Durometer hardness test.* Elastomers are polymers that behave like rubber. The Durometer hardness test is used primarily for control of consistency in polymeric materials. The test is performed by pushing a spring-loaded needle into the material. The hardness is then read from a dial or digital display. The Durometer hardness reading varies from 0 at maximum indentation (0.100 in or 2.54 mm) to 100 at zero indentation.

SUMMARY

All manufactured products are designed to provide certain services for human needs. In order to function safely, products should provide enough strength to support the expected load or mechanical force. Therefore, mechanical strength is the first factor to consider when selecting a material for design and manufacturing of any product.

Mechanical properties have been studied intensively because of the needs in design and manufacturing of products in industry. Their study started with the ideas of stress, followed by the material response to different stress conditions. Different responses of the material to the mechanical load lead to different measurements of materials properties. Among these are material strengths, modulus of elasticity, elastic and plastic behavior, ductility, fracture toughness, fatigue behavior and hardness.

IMPORTANT TERMS

Brinell Hardness Number (BHN)
Brinell Hardness Test
Brittle Fracture
Buckling
Compression Strength
Cyclic Loading
Ductile
Ductile Fracture
Durometer Hardness Test
Elastic
Endurance Curve
Endurance Limit
Extensometer
Fatigue
Fatigue Limit
Fracture Toughness
Hardness
Mechanical Property
Microhardness Tester
Modulus of Elasticity
Necking
Normal Stress

Percent Elongation
Percent Reduction in Area
Permanent Strain
Plastic Deformation
Plasticity
Residual Strain
Rockwell Hardness Test
Rockwell Superficial
    Hardness Test
Shear Modulus of Elasticity
Shear Strain
Shear Strength
Shear Stress
S-N Diagram
Static Loads
Strain
Strain Hardening
Torsion
Ultimate Tensile Strength
Yield
Yield Point
Yield Strength

REVIEW QUESTIONS

1. What is normal stress? What is the unit for normal stress in the US Customary (inch-pound) system of measurement?
2. What is normal strain?
3. What are shear stress and shear strain?
4. How is modulus of elasticity (E) calculated from a stress-strain curve?
5. What is elastic modulus of a material?
6. What is the yield point or yield strength of a material?
7. What is the ultimate tensile strength of a material?
8. Which strength property—yield strength or ultimate tensile strength—is most commonly used as a design criterion? Why?
9. What is compressive strength? Where is it used?
10. What is ductility? How is ductility applied to material selection?
11. Compare percent elongation and percent reduction in area.
12. What is fracture toughness? Why is it significant in designing aerospace structures?
13. What is fatigue? Where does fatigue usually start?
14. What is the most common way of improving fatigue resistance for a material?
15. Why is hardness so widely used as a criterion in industrial applications?
16. Compare the Brinell hardness test and a microhardness test.
17. What is Durometer hardness test? What materials are commonly evaluated with this test?
18. What is Rockwell hardness test? What does it measure?
19. Compare the Brinell and Rockwell hardness tests.

FURTHER READINGS

1. Gere, J. M. and Timoshenko, S. P. Mechanics of Materials (3rd edition). Boston: PWS - KENT Publishing Co. (1990).
2. Bolton, W. Engineering Materials Technolog. Oxford, Great Britain: Heinemann Newnes (1989).

INTERNET RESOURCES

http://www.instron.com/index1.html
Instron, a manufacturer of materials testing systems.

http://www.mts.com
MTS Systems Corp., a manufacturer of materials testing and simulation systems.

http://www.astm.org
American Society for Testing and Materials, materials property and testing standards.

Chapter 4

Physical and Chemical Properties of Materials

KEY CONCEPTS

Upon completion of this chapter, you should understand:

➢ Thermal resistance of materials.
➢ Thermal conductivity of materials and its applications.
➢ Heat capacity of materials.
➢ Thermal expansion of materials.
➢ Thermoelectrics and thermocouples.
➢ Electrical resistivity and conductivity.
➢ Dielectric constant and dielectric strength.
➢ Soft and hard magnets.
➢ Luminescence, phosphorescence, laser, and photoelectric effects.
➢ Optical reflection and transmission.
➢ Refraction.
➢ Corrosion and oxidation.
➢ Intergranular corrosion and stress corrosion cracking.

Generally speaking, physical properties of a material are the responses of the material to environmental variables. These variables include mechanical force, temperature, electromagnetic force, and so on. Physical properties can be subdivided into many categories, such as mechanical, thermal, magnetic, electrical, chemical, and so on. However, in industrial applications, mechanical properties are usually separated from physical and chemical properties. Mechanical properties of materials are discussed in Chapter 3. In this chapter, physical and chemical properties of materials are discussed, including thermal, electrical, magnetic, optical, and chemical properties. Understanding of these properties is essential for proper material selection and product development in industry.

THERMAL PROPERTIES

Thermal properties of a material are the responses of the material to changes in temperature. For example, a material's mechanical properties may vary with temperature. Heat conductivity is an important factor in choosing insulator for a house, heat-sealing heads, heat exchangers, heat sinks in electronic devices, and even spacecraft. Thermal expansion of materials has to be considered in the construction industry, as well as in the design and use of a precision instrument.

The thermal properties that need to be considered when selecting a material are heat resistance, thermal conductivity, heat capacity, and thermal expansion. Another important thermal property is thermoelectric behavior. This property is related to thermocouple applications used for temperature measurement.

TEMPERATURE DEPENDENT PROPERTIES

Many critical industrial applications depend on a combination of mechanical properties exhibited at elevated temperature or at low temperature. For example, the strength of industrial materials decreases with increasing temperature. Some temperature-critical applications include aircraft engines, cutting tools for machining, power generators, pressure vessels, and rocket components. *Heat resistance* refers to the ability of a material to maintain its desired mechanical strength at elevated temperature. Materials that can maintain their mechanical properties at elevated temperature are called *heat-resistant materials.*

Ceramics are among the most heat-resistant materials. They are suitable for applications such as crucible bricks, refractory bricks, and ceramic internal combustion engines. Metals are second to ceramics in heat resistance. Polymers are not typically used at high temperatures. The maximum application temperature for ceramics can be as high as 5000°F (2760°C). Metals can be used at temperatures up to 1500°F (815°C). Polymers, on the other hand, are limited to uses below 500°F (260°C).

At low temperatures, materials tend to become brittle and less ductile. Metals with body center cubic (bcc) crystal structure, polymers, and ceramics have a transition temperature. The *transition temperature* is the point below which a material loses its ductility. Above the transition temperature, the material is ductile. Below the transition temperature, the material becomes brittle. See Figure 4-1. The transition temperature varies for different materials. For metals and polymers, it is between −200°F and 200°F (−130°C and 95°C). For ceramics, the transition temperature is above 1000°F (540°C).

THERMAL CONDUCTIVITY

Thermal conductivity is the ability of a material to transfer heat from a body of higher temperature to a body of lower temperature. The higher the thermal conductivity of a material, the faster heat can be conducted through it. The thermal conductivity of a material is determined by its internal structure. Alloying and impurities usually decrease the thermal conductivity. Figure 4-2 lists the thermal conductivity of several typical industrial materials at room temperature of 70°F (20°C).

Metals have better thermal conductivity than polymers and most ceramics. Silver is the metal with the best thermal conductivity. However, copper is a practical choice for heat exchangers and automobile radiators because of its good thermal conductivity, low cost, and ease of forming.

Figure 4-1.
Above their transition
temperature, materials are
ductile. Below their transition
temperature, they are brittle.

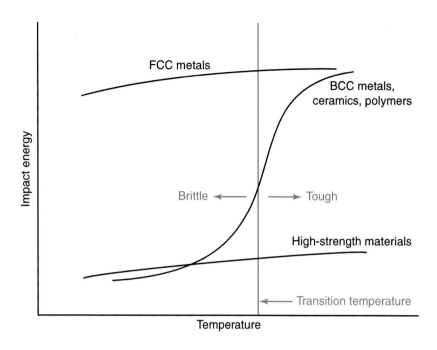

Figure 4-2.
Thermal conductivity varies
by material.

THERMAL CONDUCTIVITY (k)	
Material	**Thermal conductivity (W m-1 K-1)**
Metals	
Silver	410
Copper	390
Aluminum	200
Mild steel	54
Stainless steel	16
Polymers	
Bakelite® (thermoset)	0.23
Wood	0.08
Nylon 66 (thermoplastic)	0.025
PVC (thermoplastic)	0.0019
Ceramics	
SiC	490
Alumina (Al_2O_3)	2
Glass	0.8

Most polymers are not good thermal conductors. Therefore, they are
commonly used as insulation, such as in winter jackets. Porous materials,
such as textiles and foamed plastics, have low thermal conductivity. They
are good insulators and widely used for energy conservation.

Ceramics have a wide range of thermal conductivity. Glass has a very low thermal conductivity and is good for building insulation, such as fiberglass. On the other hand, SiC ceramic has even better thermal conductivity than silver.

Figure 4-3 shows a large furnace used to cast 138″ (3.5 m) diameter mirrors. The furnace is lined with heaters insulated by fibrous ceramics. The ceramic insulator must have low conductivity. However, it must also have a high mechanical strength because the furnace is rotated.

HEAT CAPACITY AND THERMAL EXPANSION

The efficiency of a material to absorb thermal energy is known as *heat capacity.* It is measured as the heat needed to raise the material one degree. It is usually measured as the number of calories needed to raise one gram of a material one degree Celsius (or Kelvin). Typical heat capacities are 1 cal g^{-1} K^{-1} for water, 0.11 cal g^{-1} K^{-1} for iron, 0.14 cal g^{-1} K^{-1} for glass, and 0.54 cal g^{-1} K^{-1} for thermoplastic polyethylene. As an example, the heat capacity value can be used to estimate the energy required to heat a piece of steel to certain temperature for heat treatment.

Nearly all materials expand when heated and contract when cooled. The expansion and contraction of a material is calculated:

$$\frac{\Delta L}{L_o} = \alpha \Delta T$$

where L_o is the original length and ΔL is the change in length caused by the change in temperature (ΔT). Greek letter α (alpha) is known as the *coefficient of linear thermal expansion.* Figure 4-4 lists the coefficient of linear thermal expansion at room temperature for common materials.

Figure 4-3.
The 6 m diameter furnace used to spincast 3.5 m diameter telescope mirrors at the University of Arizona Mirror Laboratory is lined with ceramic fiber insulated heaters.
(ZIRCAR Ceramics, Inc.)

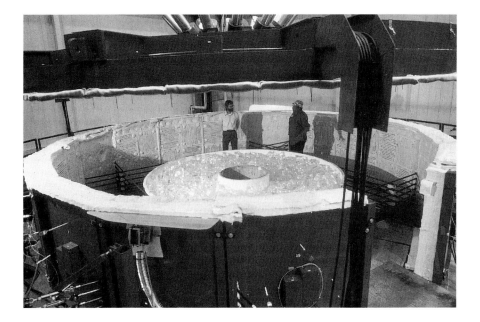

Figure 4-4.
The coefficient of linear thermal expansion for various materials.

COEFFICIENT OF LINEAR THERMAL EXPANSION	
Material	**Coefficient of thermal expansion at room temperature, α in 10^{-5} K^{-1}**
Hard rubber	8
Aluminum	2.39
Copper	1.67
Brass	1.8
Iron	1.23
Glass (ordinary)	0.9
Glass (Pyrex)	0.32
Invar alloy	0.07
Quartz glass	0.05

Figure 4-5.
A larger thermal expansion of a steel sheet causes tensile stress in the joined ceramic strip.

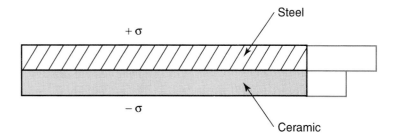

Thermal expansion of materials is an important factor in the design and selection of materials for structures. Different thermal expansions due to different materials can cause stress in a component. This is even more important with ceramics since ceramic components cannot undergo plastic deformation to allow for movement. For example, suppose a ceramic plate is joined with a strip of steel. When heated, the steel expands more than the ceramic, as shown in Figure 4-5. This causes tension stress on the ceramic part. The tension stress promotes cracking in the ceramic and may cause a fracture.

A bimetallic thermostat is a perfect example of where the difference in thermal expansion of different materials is used. The thermostat consists of two metallic strips welded together. One metal has a low coefficient of thermal expansion, such as Invar alloy ($\alpha = 7 \times 10^{-7}$ K^{-1}). The other metal has a high coefficient of thermal expansion, such as brass ($\alpha = 1.80 \times 10^{-5}$ K^{-1}). The beam is straight when the ambient temperature is the same as welding temperature. See Figure 4-6. However, when the temperature of the beam changes, the two metals expand (or contract) by different amounts. This is because of the difference in coefficient of thermal expansions between the two metals. The difference in expansion (or contraction) between the top and bottom layers causes the beam to deform into an arc. When a switch is attached to the beam, another device, such as a furnace, can be activated on temperature change.

Figure 4-6.
A bimetallic strip is straight at the welding temperature (T_w) and curved when temperature changes.

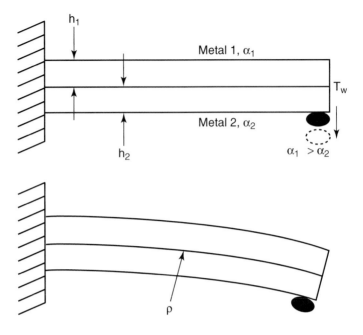

THERMOELECTRICS

When two dissimilar metals are brought into contact, a voltage is present across the junction. The amount of voltage depends on the chemical composition of the metals and the temperature. This is a thermoelectric behavior known as the *Peltier effect.*

If there is a temperature difference between the ends of a wire made from a single metal, a voltage is present between the ends of the wire. The amount of voltage is determined by the wire material, the chemical uniformity of the material, and the temperature difference. This thermoelectric behavior of material is known as the *Thomson effect.*

Thermoelectric behavior of materials is the principle behind a thermocouple. This device is commonly used in many industries. See Figure 4-7. The total voltage depends on the temperature difference between the two junctions. This thermoelectric phenomenon was discovered in 1821 by T. J. Seebeck and is called *Seebeck effect.* An application of a thermocouple is a precise thermometer. A voltmeter is used to measure the voltage produced by the thermocouple and the temperature is determined from the voltage.

ELECTRICAL PROPERTIES

Electrical properties include resistivity, conductivity, dielectric properties, semiconductivity, and superconductivity. Electrical properties of a material are important in just about every industry, including those involved in electrical power transmission and distribution, computers, television components, and lasers. Understanding a material's electrical properties is critical for success in the design, manufacture, and quality control of electronic components and systems.

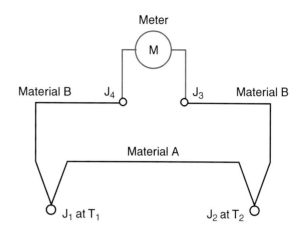

RESISTIVITY

The *resistivity* of a material is a measure of how well the material resists the flow of electricity. Electrons move inside a conductor to produce a current under the influence of a voltage. The amount of electrical current (I) is directly proportional to the voltage (V) and inversely proportional to the electrical resistance of the conductor (R).

$$I = \frac{V}{R}$$

This relationship is known as Ohm's law. The resistance of a material is constant based on temperature and specimen size.

The electrical resistance can be altered by changing the size of the conductor. Resistance increases as the length of a conductor increases. Resistance also increases as the cross-sectional area of a conductor decreases. The relationship can be expressed mathematically as:

$$R = \rho \frac{L}{A}$$

where L is the length, A the cross-sectional area, and ρ (Greek letter rho) is the electrical resistivity of the material. For metals, when the temperature increases, the resistivity increases. In other words, they have positive temperature coefficients of resistivity. However, for semiconductors, such as Germanium (Ge) and Silicon (Si), an increase in temperature leads to a decrease in the resistivity. They have negative temperature coefficients of resistivity. Figure 4-8 lists the resistivity and temperature coefficient for some typical materials.

For conducting electricity, a material with lower resistivity should be selected. This is because the resistance is related to the power consumption and heat generation. Gold and silver are used in the electrical contacts because of their low electrical resistivity and high resistance to oxidation. A material of high resistivity is used as an insulator. Ideally, the material used as a resistor should have a very low temperature coefficient of resistivity so the resistivity will not change significantly with a variation in temperature.

Figure 4-8.
Resistivity and temperature
coefficient of resistivity for
various materials.

RESISTIVITY AND TEMPERATURE COEFFICIENT OF RESISTIVITY		
Material	Resistivity ($\Omega \cdot m$)	Temperature coefficient α [C^{-1}] at 20°C
Silver	1.59×10^{-8}	3.8×10^{-3}
Copper	1.70×10^{-8}	3.9×10^{-3}
Gold	2.44×10^{-8}	3.4×10^{-3}
Aluminum	2.82×10^{-8}	3.9×10^{-3}
Tungsten	5.60×10^{-8}	4.5×10^{-3}
Iron	10×10^{-8}	5.0×10^{-3}
Platinum	11×10^{-8}	3.92×10^{-3}
Lead	22×10^{-8}	3.9×10^{-3}
Nichrome	150×10^{-8}	0.4×10^{-3}
Carbon	3.5×10^{-8}	-0.5×10^{-3}
Germanium	0.46	-48×10^{-3}
Silicon	640	-75×10^{-3}
Glass	$10^{10} - 10^{14}$	
Hard rubber	$10^{13} - 10^{6}$	
Sulfur	1015	
Quartz (fused)	75×10^{16}	

CONDUCTIVITY

The *conductivity* of a material is a measure of how well the material conducts electricity. Conductivity is the reciprocal of its resistivity and represented by the Greek letter σ (sigma), expressed as:

$$\sigma = \frac{1}{\rho}$$

where σ is the resistivity of material.

Conductivity is associated with the particles existing inside a given material. For instance, metals typically have many free electrons. This leads to characteristically high conductivity for metals. Electrons also account for the conductivity of n-type semiconductors. However, the low concentration of electrons in these materials results in a lower conductivity. In an ionic compound, such as Al_2O_3, the concentration of free electrons is so low that ions determine the conductivity. As a result, the conductivity is very low and these materials are commonly used as insulators.

DIELECTRIC PROPERTIES

A *dielectric* is a nonconducting material, such as rubber or glass, inserted between two plates of a capacitor. Dielectric properties are of major concerns when selecting insulation for electrical equipment. This section discusses dielectric constant and dielectric strength. These properties are commonly used in evaluating insulating materials.

Capacitance is the ability of a material or device to store electrical charges. The capacitance of a parallel plate capacitor with vacuum space as the dielectric is proportional to the area of the plates and inversely proportional to the distance between the plates. This is expressed as:

$$C = \varepsilon_o \frac{A}{d}$$

where C is the capacitance, A is the area of the plates, d the thickness of the dielectric, and ε_o is the permittivity constant of vacuum space. Permittivity is the ability of a dielectric to store electrical potential.

If a dielectric material, such as mica, is placed between the plates of the same capacitor, the capacitance is increased by a factor of κ (Greek letter kappa). This factor is known as *relative permittivity* or *dielectric constant.* The dielectric constant measures the charge-storing capacity of a dielectric in a capacitor in relation to vacuum:

$$\kappa = \frac{\varepsilon}{\varepsilon_o}$$

where ε is the permittivity of the dielectric material and ε_o is the permittivity of a vacuum. From this equation, you can see that the dielectric constant of a vacuum is one. The dielectric constant of a material depends on the ability of the material to react and orient itself to the electrical field applied. The greater the reaction, the greater the energy stored, therefore, the higher the dielectric constant of the material.

The *dielectric strength* of an insulating material is the limiting electric field intensity above which an electrical breakdown occurs. When the breakdown happens, electrical conductivity is induced in the material and it is no longer able to store charges. This dielectric strength is commonly expressed in voltage gradient. For example, volts per meter (V/m) is a common unit designation. Figure 4-9 lists the dielectric constants and

Figure 4-9.
Dielectric constant and
strength of various materials.

DIELECTRIC CONSTANTS AND DIELECTRIC STRENGTHS AT ROOM TEMPERATURE		
Material	**Dielectric constant (κ)**	**Dielectric strength (V/m)**
Vacuum	1	—
Air	1.00059	3×10^6
Bakelite	4.9	24×10^6
Fused quartz	3.78	8×10^6
Pyrex glass	5.6	14×10^6
Polystyrene	2.56	24×10^6
Teflon®	2.1	60×10^6
Neoprene rubber	6.7	12×10^6
Nylon	3.4	14×10^6
Paper	3.7	16×10^6
Strontium titanate	233	8×10^6
Water	80	—
Silicone oil	2.5	15×10^6

dielectric strengths of various materials at room temperature. Notice that most insulating materials have a dielectric strength and a dielectric constant greater than those of air.

Dielectric strength varies with the thickness of the insulating material, moisture absorption, and the length of time and frequency in which the voltage is applied. An increase in thickness increases the necessary breakdown voltage, but not in direct proportion. Moisture, contamination, elevated temperature, aging, and mechanical stress usually decrease the dielectric strength of the material. Therefore, testing conditions should follow those specified by American Society for Testing and Materials (ASTM) standards.

MAGNETIC PROPERTIES

Magnetism is the mutual attraction of two iron-based materials. The phenomenon of magnetism was known to the ancient world. A compass invented by the Chinese was the first application of magnetic material. This device made an important contribution to navigation and the discoveries of unknown lands. Today, various magnetic materials impact our daily lives. Tape recorders, television sets, computers, electric motors, and particle accelerators are just a few applications that use magnetic materials.

If a magnetic field with field strength H is applied across a material, there is a magnetic flux with density B through the gap. The ratio of field strength to flux density (H/B) is known as the *permeability* (μ) of the material. The important difference between the electric and magnetic circuits is that the conductivity in an electric circuit is constant. However, the permeability of a magnetic material changes with the applied field strength. The ratio H/B is not constant. Also, the flux can persist after the magnetic field is removed. This is called **remnant induction.**

A typical relationship between the flux density and field strength plotted as curve is shown in Figure 4-10. When the material is magnetized the

Figure 4-10.
A complete cycle of magnetization forms a hysteresis loop.

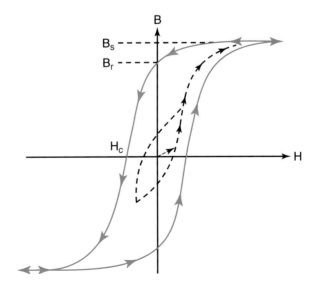

first time, the flux density is increased following the dashed curve in the figure as the field strength increases. The value of B rises rapidly and then practically levels off at B_s. To lower the flux density to zero, a field strength H_c is needed in the opposite direction to the original field. This is called the *coercive magnetic force* or *coercive field.* The graph never retraces the original curve, but traces out a loop. This loop is known as a hysteresis loop. The *hysteresis loop* is an index of the lost energy in a complete cycle of magnetization.

The shape of the B-H curve varies greatly with different magnetic materials, and various types of B-H behavior have different applications. For example, a transformer core material should have a very small area inside the hysteresis loop. Since the hysteresis loop represents a power loss per cycle, it should be kept as low as possible. The material used for this purpose is called a soft magnet and is easily magnetized and demagnetized. By the contrast, in a computer disk, a sizable value of B must be easily achieved and then retained when the power is off. In this way, information can be retrieved later. For this purpose, a square loop is needed where B_s is very close to B_r.

With a permanent magnet, the power of the magnet is related to area within the hysteresis loop. A much larger loop is needed for permanent magnets. Materials with this characteristic are called *hard magnets.*

OPTICAL PROPERTIES

Optical properties are responses of materials to light. They were studied originally for the application of glass, lenses, and colors. The most visible properties of metals—luster and color—have been known since metals were discovered. Metals have been used since ancient times for mirrors and jewelry. Material color was also utilized over 4000 years ago by the Chinese as a guide to determine the composition of tinned bronze. Today, electro-optic materials are used in lasers, electron microscopes, fiber-optics, and optic information storage for computers. Optical properties include emission, absorption, reflection, transmission, and refraction.

REFLECTION

When a light shines onto a very smooth surface, such as a mirror, nearly all of the light is reflected off the material at the same angle. This reflection of light is called a *specular reflection.* On the other hand, if the reflecting surface is rough, the surface will reflect the rays to various directions. This kind of reflection is known as a *diffuse reflection.*

The ability of a material to produce a specular reflection is known as *reflectivity.* Metals typically have high reflectivity. This is because light cannot penetrate deep into a metal. In contrast, light penetrates much farther into glass, if not completely through.

White light consists of violet, blue, green, yellow, and red light. When the white light shines on different materials, different colors are seen. This is because the material reflects specific colors. For example, gold reflects primarily yellow light. However, copper absorbs blue light, thus reflecting the other light colors to produce a reddish color. Reflection color can be used to identify different materials.

EMISSION AND ABSORPTION

Emission refers to electromagnetic radiation from a material. This radiation can be a visible light, like the yellow light produced when salt powder is thrown into a fire. Radiation can also be outside the visible light range, Figure 4-11. Radiation such as X-rays or gamma rays have wavelengths shorter than those of visible light. Microwaves and radio waves have longer wavelengths than those of visible light. Emission is caused by electron energy excitation inside the material.

When light strikes a material, a portion of the light is absorbed by the material. This is called **absorption.** It is important to distinguish between color produced by light *emission* and by *absorption.* A yellow color is *emitted* when salt is thrown in a fire because of the characteristic wavelength of the sodium atom. On the other hand, when white light passes through yellow glass, yellow light is produced. This is because the glass *absorbs* all colors of light in white light except yellow.

When a light strikes certain materials, electrons are emitted from the surface. The material then becomes positively charged. This phenomenon is known as the **photoelectric effect.** The emitted electrons are called **photoelectrons.** The photoelectric effect is applied in photoelectric sensors. Materials that can produce the photoelectric effect are usually compounds of the alkali metals, such as AgOCs, SbCs, NaKSbCs, and CsTe. The most common material used to produce the photoelectric effect is SbCs.

Figure 4-11.
Characteristics of electromagnetic radiation are related to wavelength in the electromagnetic spectrum.

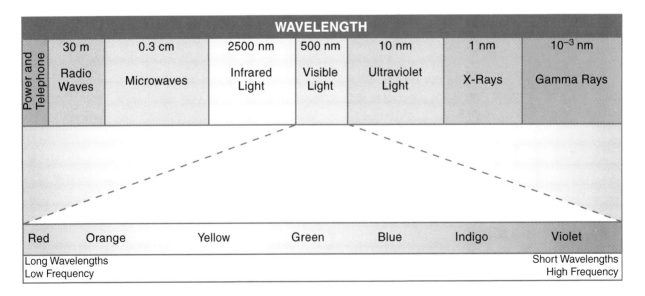

LUMINESCENCE

Certain materials emit visible light, or glow, when exposed to ultraviolet light. This property is known as *luminescence.* These materials are said to be *fluorescent.* Some luminescent materials continue to glow after the ultraviolet light is removed. This property is called *phosphorescence.*

Fluorescent and phosphorescent minerals are quite important. For example, these types of minerals are used to produce television picture tubes. The minerals selected must produce pure primary colors (red, green, and blue). They must also quickly cease to glow when ultraviolet light is removed. Otherwise, the picture will be blurry and the TV will continue to glow after turned off. The ultraviolet light is provided from an electron gun inside the TV. Phosphorescent minerals are used in road signs and glow-in-the-dark clock faces.

A laser is another example of where luminescence is important. In a ruby laser, for example, a single crystal rod of alumina (Al_2O_3) contains a small quantity of Cr^{3+} ions. The rod is ground with flat surfaces at each end. One end is made completely opaque while the other end is made only slightly opaque. A tube containing xenon gas emits a constant wavelength of light onto the tube. Stimulated electrons in Cr^{3+} ions of the rod emit an accelerated, high-powered light beam from the crystal. There are many uses for lasers, including surgical and material cutting, surveying, precision measurement, and process control.

TRANSMISSION

Metals are generally very opaque. Visible light generally cannot be transmitted through metals. In the case of ionic and covalent solids, such as glass, lower-energy visible and infrared lights penetrate into the solids and transmit through them. These solids are said to be transparent. However, very-high-energy radiation such as ultraviolet is absorbed.

If atoms of an impurity are present in a transparent material, absorption can occur. For example, in a ruby, the chromium ions absorb the blue and green lights, but allow the remaining light to pass through. This light is mostly of wavelengths corresponding to red light.

Semiconductors are a special case. They are transparent to infrared radiation. However, they appear opaque in visible light. Semiconductors can be used as windows or lenses for efficient transmission of infrared radiation.

REFRACTION

Have you ever noticed how a stick appears bent when you put it in water? This is due to light refraction. *Refraction* is the bending of a light ray. As the light ray moves from the first medium (such as air) to the second medium (such as water), it is bent. The light is said to be refracted. The angle of refraction is the angle between the refracted ray and a line perpendicular to the surface. See Figure 4-12. The angle of refraction (θ_2) is smaller than the angle of incidence (θ_1). These two angles follow the relationship:

$$\frac{Sin\theta_1}{Sin\theta_2} = constant$$

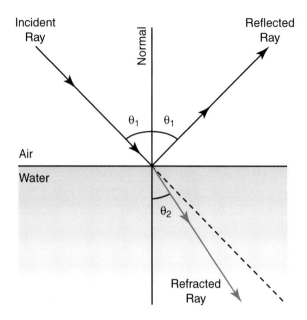

Refraction occurs because the speed of light is different in the two media. The speed of light is highest in a vacuum. An *index of refraction (n)* is defined as:

$$n = \frac{\text{speed of light in a vacuum}}{\text{speed of light in a medium}} = \frac{c}{v}$$

Air has an index of refraction close to that of a vacuum (n = 1.000293). Most solids and liquids have an index of refraction greater than 1.3.

CHEMICAL PROPERTIES

Mechanical and physical properties are important selection criteria to ensure a material resists load or stresses and performs to design functions. In many instances, the chemical properties of a material are just as important. *Chemical properties* are a measure of how a material interacts with its environment, such as gases, liquids, and solids.

Most automobile owners are concerned about rust patches appearing on the car. Rust not only makes the body look shoddy, but it also indicates a mechanical weakening of the metal. Rust indicates the steel has deteriorated due to an interaction with the environment. The resistance to corrosion is a critical consideration for normal and safe operation of equipment.

CORROSION

Corrosion is defined as the destruction or deterioration of a material due to a chemical reaction with its environment. Corrosion is usually in the form of material removal and oxide formation. Corrosion happens with metals, ceramics, and polymers. All environments are corrosive to some degree. Examples of corrosive environments include humid air, saltwater, steam, chlorine, ammonia, sulfur dioxide gas, acids, and solvents. Materials exposed to these environments must be corrosion resistant.

The corrosion resistance of a material depends on two major factors. The first factor is how chemically active the material is. For example, gold is a noble metal and barely reacts with water. However, steel quickly rusts in contact with water. The second factor is the nature of the surface film formed due to the reaction with an environment. Most metals form a natural protective film on their surface after being exposed to air for an extended period of time. Once formed, the coating prevents further corrosion. An example is aluminum. An aluminum wire reacts easily with oxygen. However, the aluminum oxide coating is not porous and further reaction is hindered. On the other hand, an iron wire reacts with air more slowly. However, the reaction continues after iron oxide forms because it is porous and continues to allow air in contact with the iron.

Corrosion takes place in many forms. *Uniform attack* or *general corrosion* is characterized by a uniform chemical or electrochemical reaction over the entire exposed surface. *Galvanic corrosion* occurs when two or more dissimilar metals are in contact, one of which is corroded much faster. *Pitting* results in cavities or pits in localized areas. The pits can range from deep, small diameter pits to shallow, large diameter pits. *Intergranular corrosion* occurs typically at the grain boundaries of stainless steel. This can result in catastrophic failure. *Stress corrosion cracking* is the result of accelerated corrosion caused by either a residual stress in the metal or an externally applied stress from a specific chemical. *Corrosion fatigue* results when the corrosion is accelerated by cyclic or repeated stress.

Stainless steel is the most commonly used material when a high corrosion resistance is needed. The surface film (chromium oxide) created on stainless steel is very nonporous and adheres tightly to the base metal. However, intergranular corrosion and stress corrosion cracking are potential problems with stainless steel. Intergranular corrosion can be caused by the improper heat treatment of stainless steel. All stainless steels contain certain amounts of carbon. When the steel is heated to its sensitizing temperature range, as when welded, the carbon is precipitated out at the grain boundaries to form chromium carbide. This formation removes chromium from the stainless steel, thus making the boundary much more susceptible to corrosion attack. The intergranular corrosion of stainless steel usually develops without being noticed. Therefore, proper treatment of stainless steel is important to avoid intergranular corrosion. An effective way to eliminate intergranular corrosion is to use super-low-carbon stainless steel.

A material may fail at a much lower stress than its yield strength if a specific chemical presents. This type of failure is caused by *stress corrosion cracking.* In stress corrosion cracking, virtually no corrosion appears over most of a material's surface. However, fine cracks develop through the material. This cracking can cause failure at stresses within the design range without any warning. Fairly specific material-chemical medium conditions are necessary for stress corrosion cracking to occur. For example, stainless steel develops stress corrosion cracks in chloride environments, but not in an ammonia environment. Brass, on the other hand, develops stress corrosion cracks in an ammonia environment.

OXIDATION

Oxidation is the reaction between a metal and oxygen without water present. Oxidation is really a special form of corrosion. Nearly every metal and alloy reacts with oxygen at elevated temperatures. Therefore, the oxidation resistance of a material is important in high-temperature applications. Some high-temperature applications include gas turbines, rocket engines, furnaces, and high-temperature petrochemical systems.

When exposed to high temperature, metals develop a thick oxide film known as *scale.* Scale-resistant materials are needed between 800°F – 2000°F (430°C – 1095°C). Adding chromium, silicon, and aluminum to iron leads to the formation of a protective film at high temperatures. This surface film adheres to the iron and prevents further oxidation of the material. For example, gray cast iron with 2% silicon has an oxidation resistance that makes it suitable for high-temperature applications.

Some gases can combine with the normal metal scale to create a low-melting-point liquid in the scale at high temperature. This liquid provides rapid transport of oxygen to the metal surface and, therefore, oxidation is accelerated. This accelerated attack is usually known as *catastrophic oxidation.*

Another type of oxidation is *internal oxidation.* In this case, oxygen dissolves in the metal, migrates inward, and forms oxide particles inside the material. Most of the time the oxygen tends to react first with the alloying elements inside. This causes some hardening, but the resulting brittleness is usually undesirable in high-temperature alloys.

SUMMARY

Physical properties of materials include thermal, electrical, magnetic, optical, and chemical properties. Each of these properties is important for specific material applications. For example, thermal properties of a material are important in high-temperature or low-temperature applications. Chemical properties are important in highly corrosive environments. When selecting a material, the physical properties must match the needs of the application.

KEY TERMS

Absorption	Internal Oxidation
Catastrophic Oxidation	Luminescence
Chemical Properties	Magnetism
Coefficient of Linear	Oxidation
Thermal Expansion	Peltier Effect
Coercive Field	Permeability
Coercive Magnetic Force	Phosphorescence
Conductivity	Photoelectric Effect
Corrosion	Photoelectrons
Corrosion Fatigue	Pitting
Dielectric	Reflectivity
Dielectric Constant	Refraction
Diffuse Reflection	Relative Permittivity
Emission	Remnant Induction
Fluorescent	Resistivity
Galvanic Corrosion	Scale
General Corrosion	Seebeck Effect
Hard Magnets	Specular Reflection
Heat Capacity	Stress Corrosion Cracking
Heat Resistance	Thermal Conductivity
Heat-Resistant Materials	Thermal Properties
Hysteresis Loop	Thomson Effect
Index of Refraction	Transition Temperature
Intergranular Corrosion	Uniform Attack

REVIEW QUESTIONS

1. What are physical properties of materials? Why are they important to modern industry?

2. What is a heat-resistant material?

3. Suppose one side of a bimetallic strip has a coefficient of thermal expansion (α) of $7 \times 10^{-7}\,\text{K}^{-1}$. The other side has a α of $1.80 \times 10^{-5}\,\text{K}^{-1}$. What will happen when the strip is heated?

4. What is thermocouple? How can it be used to measure temperature?

5. What is a dielectric constant?

6. To reduce the area of the capacitor plates but maintain the same capacitance, a new material with a _____ dielectric constant should be selected.

7. Early capacitors used paper as their dielectric. Assume the thickness of the paper is 2 mm and the dielectric strength of paper is $16 \times 10^6\,\text{V/m}$. What is the maximum voltage that can be applied without leading to any electrical breakdown?

8. Why does an electrical transformer have a soft magnet as its core material?

9. What are fluorescence and phosphorescence?

10. What are typical applications for fluorescence and phosphorescence?

11. What is photoelectric effect?

12. Why are semiconductors used to construct infrared sensors?

13. Why does a stick appear bent in water?

14. What is corrosion? Why is it important in material selection?

15. What is the difference between corrosion and oxidation?

16. What is intergranular corrosion?

17. In what type of material does intergranular corrosion typically occur?

18. What is stress corrosion cracking?

19. What is internal oxidation?

20. What is catastrophic oxidation?

FURTHER READINGS

1. Flinn R.A. and Trojan P.K. Engineering Materials and Their Applications (4th edition). Boston: Houghton Mifflin Co. (1990).

2. Hummel R. E. Electronic Properties of Materials (2nd edition). Berlin: Springer-Verlag (1992).

3. Fontana, M.G. Corrosion Engineering (3rd edition). New York: McGraw-Hill Book Co. (1985).

4. Schweitzer P.A. What Every Engineer Should Know about Corrosion. Marcel Dekker, Inc. (1987).

5. Serway R.A. Physics for Scientists and Engineers with Modern Physics (4th edition). Philadelphia: Saunders College Publishing (1997).

6. Pollock D. D. Physical Properties of Materials for Engineers. Boca Raton, Florida: CRC Press, Inc. (1993).

INTERNET RESOURCES

http://www.ornl.gov
> Oak Ridge National Laboratory, materials science research and development

http://www.mrs.org
> Materials Research Society: Brings together scientists, engineers and research managers to share R&D findings about new materials, mainly electronic or functional applications.

http://physics.nist.gov
> Physics Laboratory, National Institute of Standards and Technology

http://dir.yahoo.com/Science/Engineering/Material_Science
> Yahoo's materials science index

http://www.copper.org
> Copper Development Association, Inc.

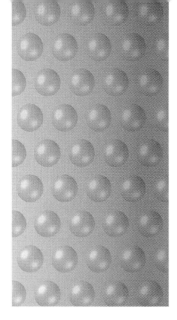

Chapter 5

Ferrous Alloys

KEY CONCEPTS

Upon completion of this chapter, you should understand:

➢ Definitions and main applications of steel and cast iron.
➢ The basic compositions of plain-carbon steel, alloy steel, high-alloy steel, and cast iron.
➢ The basic process of making iron and steel.
➢ Iron-iron carbide phase diagram and equilibrium structures of steel.
➢ Application examples of plain-carbon steels, alloy steels, and cast irons.
➢ Application of stainless steels.
➢ Heat treatment of steel and martensitic strengthening.
➢ Surface hardening and its applications.

Ferrous alloys generally refer to iron-based alloys, or iron and steel. Since iron was discovered and first used approximately three thousand years ago, ferrous alloys have remained important. Today, many applications of iron and steel can be found in everyday life. Automobiles, tools, and buildings all rely on steel for their manufacturing or construction. Approximately 85% of all metals used for industrial applications in the United States are iron based. Surprisingly, however, this material so common today has been in large-scale commercial use for only about 150 years.

Though the use of nonferrous alloys, polymers, and composites is increasing, iron and steel are still used for those applications that require a heavy load bearing capacity. The popularity of iron and steel is principally due to:

♦ Low cost.
♦ Wide range of attainable mechanical properties.
♦ High modulus of elasticity.

This chapter discusses the composition and production of ferrous alloys, structural change through a phase diagram, characteristics of steel and cast iron, and heat treatment of iron and steel.

COMPOSITION

Ferrous alloys are basically iron with a certain amount of carbon added. *Steel* usually contains less than 1.7 percent of carbon. *Cast iron* normally has 2 to 4 percent of carbon. There are other elements also present in iron and steel. Silicon and manganese are often added because they improve the mechanical properties of the alloy. These are called *alloying elements.* Sulfur and phosphorus are detrimental to the mechanical properties of iron and steel. These are called *impurities.*

Plain carbon steels are defined as alloys of only iron and carbon. Other elements are not present in sufficient quantities to affect the structure and properties. When enough alloying elements are present in plain carbon steel to affect the structure and properties, but less than 5 percent total, the material is called an *alloy steel*. When large amounts of alloy elements are added, the steel is called *high-alloy steel*. For example, stainless steel consists of 18 percent chromium and 8 percent nickel in a low-carbon steel. This steel has a high corrosion resistance. Large amounts of chromium, tungsten, vanadium, and other alloy elements are added to plain carbon steel to make *tool steel.* This is a very hard steel used to make cutting tools, dies, and other tools. *Superalloys* are plain carbon steel with more than 50 percent of alloy elements. These materials are used for special purposes, such as high-temperature applications.

Cast iron has much higher carbon content than steels. Normally, it also has higher concentration of silicon. The high content of carbon and silicon make cast iron hard to deform. Therefore it is best "cast" into a mold. The carbon in cast iron is normally graphite. However, white cast iron contains carbon in the form of iron carbide or cementite (Fe_3C).

Elements such as manganese, nickel, chromium, and copper, are commonly added to cast iron to create special purpose cast irons. These special purpose cast irons are utilized for corrosive, abrasive, or high-temperature applications. For example, cast iron containing 2.4–3.4 percent chromium and 3.35–4.25 percent nickel is abrasion resistant. This type of cast iron is valuable in materials handling applications, such as conveyors and impurity pumps.

CREATING IRON AND STEEL

The first step in iron and steel making is to convert iron ore into pig iron. *Iron ore* is a raw material mined from the ground that contains iron oxide (Fe_2O_3). Iron ore is converted to *pig iron* in a blast furnace, Figure 5-1. Pig iron is about 4 percent carbon. The following chemical reaction takes place inside the blast furnace.

$$Fe_2O_3 + 3CO = 2Fe + 3CO_2$$

The iron produced from the chemical reaction is liquid, or molten, at the high temperature inside the blast furnace. The molten iron then absorbs carbon as it passes through *coke* to the bottom of the blast furnace. Molten pig iron then comes out of the furnace. This pig iron can be cast or used for making steel.

Figure 5-1.
Modern blast furnaces make iron from iron ore around the clock. (Bethlehem Steel)

STEEL

Since most of the steels used today contain less than 1 percent carbon, the excess carbon in pig iron must be removed to convert it into steel. One of the techniques to achieve this conversion is to use the basic oxygen furnace. A basic oxygen furnace is a closed-bottom, refractory-lined vessel, Figure 5-2. Molten pig iron, recycled steel scrap, lime, and fluorspar are added to the furnace. The lime and fluorspar create slag. A high-velocity stream of oxygen is directed down onto the molten mixture through a water-cooled lance. This causes rapid oxidation of carbon, thus lowering the carbon content. Deoxidizers and any required alloying elements are added as the steel is removed from the furnace. The molten steel is then poured into ingot molds or a continuous casting line.

Figure 5-2.
A high-velocity stream of oxygen is directed down onto the molten iron through a water-cooled lance, causing the rapid oxidation of carbon.

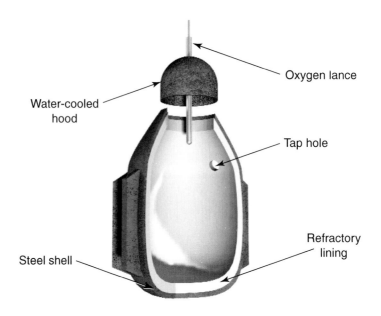

Oxygen lance

Water-cooled hood

Tap hole

Refractory lining

Steel shell

Steel can be processed into various forms. Figure 5-3 shows a typical steel tube production process. For example, molten steel can be cast into steel rods using a continuous casting process. The rods are then rolled to the desired diameter before being identified and cut to final length. Then, the steel rods are fabricated into a hollow tube by piercing mill, elongator, reducing mill, and rotary sizer. The tubes are subject to different thermal and mechanical treatments in order to obtain desired mechanical properties before shipping.

CAST IRON

Cast iron is converted from pig iron in a smelting furnace. Most cast iron foundries in the United States use electric induction furnaces. Pig iron, recycled scrap iron and scrap steel, and limestone are added to the furnace. Unlike with steel, accelerated oxidation of carbon is not induced. However, alloying elements are added to the molten metal. The molten metal is then removed from the furnace and cast into various shapes.

IRON-IRON CARBIDE (Fe–Fe$_3$C) PHASE DIAGRAM

The properties of iron and steel are determined by their microstructure. In order to control the microstructure, it is essential to understand the *iron-iron carbide (Fe–Fe$_3$C) phase diagram*, Figure 5-4. A pure iron has a body centered cubic (BCC) crystal structure at room temperature. This is called **α iron.** When the α iron is heated to 1674°F (912°C), it is transformed to face-centered cubic (FCC) **γ iron.** This is then transformed to body-centered cubic **δ iron** at 2541°F (1394°C).

$$1674°F\ (912°C) \qquad\qquad 2541°F\ (1394°C)$$
$$\text{α iron (BCC)} \longrightarrow \text{γ iron (FCC)} \longrightarrow \text{δ iron (BCC)}$$

Figure 5-3.
A steel tube is produced through continuous casting, rolling, piercing, and finished treatment. (The Timken Company)

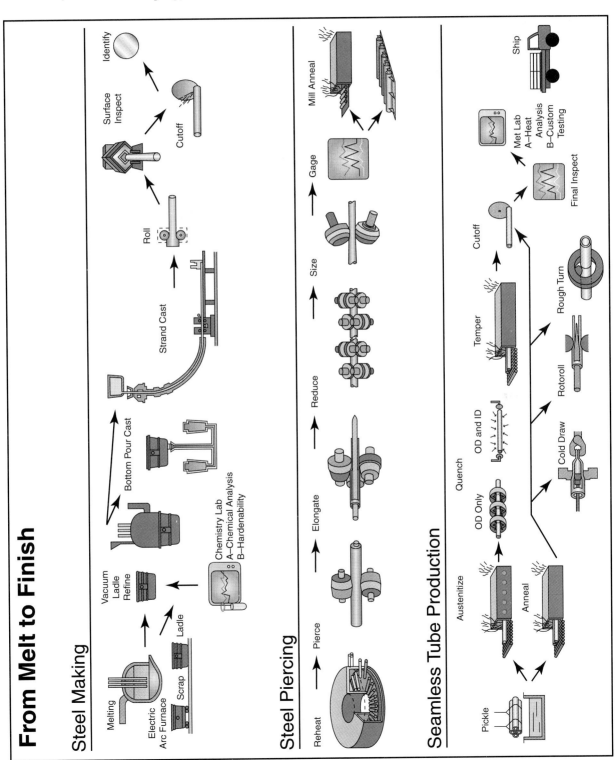

Figure 5-4.
A iron-iron carbide phase
diagram demonstrates
various possible phases
present in iron and steel at
different compositions and
temperatures.

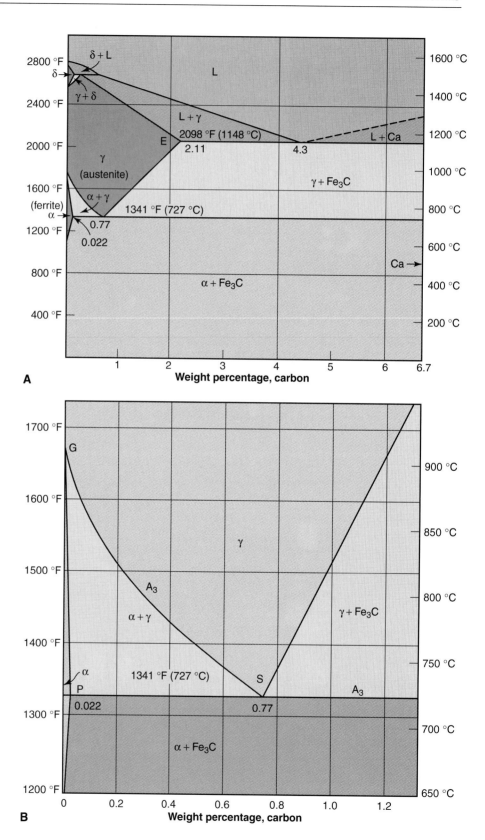

FERRITE (α)

When a trace amount of carbon is added to pure iron (or α iron) at low temperature, carbon atoms are evenly distributed in the spaces in-between iron atoms. This microstructure is known as α ferrite. The properties of α ferrite are similar to pure iron. It is very soft and can be easily deformed.

As temperature increases, the amount of carbon that can be dissolved into iron increases slightly. As shown in a phase diagram, only 0.005 percent carbon can be dissolved into pure iron at 32°F (0°C). The maximum amount of carbon is 0.02 percent at 1341°F (723°C).

AUSTENITE (γ)

When carbon atoms are dissolved into γ iron at high temperature, they form a microstructure called austenite. Austenite has a face centered cubic (FCC) crystal structure. More carbon can be dissolved into it than into ferrite. A maximum of 2.08 percent carbon can be incorporated into iron at 2098°F (1148°C). This is 100 times more than ferrite. Austenite is also ductile and easy to deform.

CEMENTITE (Fe_3C)

Besides being dissolved into iron, carbon atoms can also form a compound with iron atoms. The compound of iron and carbon is called cementite (Fe_3C). It has a composition of 6.67 percent carbon and 93.3 percent iron. Cementite is very hard and brittle.

A phase diagram shows the phases present at certain temperatures and compositions. For example, a steel with 0.5 percent carbon has a single phase of austenite at 2000°F (1095°C). When cooled to 800°F (430°C), the steel has ferrite plus cementite (Fe_3C).

STEELS

Steel is a ferrous alloy that usually contains less than 1.7 percent of carbon. Other elements may also be present. Steel can be classified as plain carbon steel, alloy steel, and high-alloy steel. This section discusses plain carbon steels and alloy steels in detail.

PLAIN CARBON STEELS

Plain carbon steel contains a maximum of 1.5 percent carbon. The higher the carbon content, the harder and stronger the steel. All other alloying elements are present in small percentages. Manganese is 1.65 percent maximum, silicon 0.60 percent maximum, copper 0.60 percent maximum, and sulfur and phosphorus 0.05 percent maximum. Carbon is the principal hardening agent and significantly affects the crystal structure and strength of the steel. Thus, different plain carbon steels are designated according to their carbon contents. Figure 5-5 lists the carbon and manganese compositions of some selected grades of plain carbon steels designated by American Iron and Steel Institute (AISI).

Figure 5-5.
This table shows the
composition of plain carbon
steels.

COMPOSITION OF PLAIN CARBON STEELS		
AISI No.	**%C**	**%Mn**
1006	0.08 max.	0.25–0.40
1010	0.08–0.13	0.30–0.60
1015	0.13–0.18	0.30–0.60
1020	0.18–0.23	0.30–0.60
1025	0.22–0.28	0.30–0.60
1030	0.28–0.34	0.60–0.90
1035	0.32–0.38	0.60–0.90
1040	0.37–0.44	0.60–0.90
1045	0.43–0.50	0.60–0.90
1050	0.48–0.55	0.60–0.90
1055	0.50–0.60	0.60–0.90
1065	0.60–0.70	0.60–0.90
1070	0.65–0.75	0.60–0.90
1075	0.70–0.80	0.40–0.70
1080	0.75–0.88	0.60–0.90
1085	0.80–0.93	0.70–1.00
1090	0.85–0.98	0.60–0.90
1095	0.90–1.03	0.30–0.50
P, 0.040 maximum; S, 0.05 maximum		

The steel codes are usually called AISI codes. A four-digit number is used for the AISI code. The first two digits designate the type of steel, 10 for plain carbon steels. The second two digits show the carbon content in hundredths of percent. For example, designation AISI 1045 indicates a plain carbon steel with a nominal carbon content of 0.45 percent.

Plain carbon steels are often classified into three main categories according to their carbon contents. These categories are mild or low-carbon, medium-carbon, and high-carbon steels.

Mild Steel or Low-Carbon Steel

Low-carbon steel contains a maximum of 0.25 percent carbon. It is easily machined, formed, and welded. However, normal heat treatment cannot harden the steel significantly. Low-carbon steels are used for construction applications. They are often forged or rolled to form angle, channel, plate, bar, or rod stock. Low-carbon steels are produced in large quantity and are usually inexpensive.

Medium-Carbon Steel

Medium-carbon steel has a carbon content between 0.25 percent and 0.65 percent. The hardness and strength of these steels can be improved by heat treatment. Medium-carbon steels are the most versatile of all plain carbon steels and used for a wide range of applications. For example, many automobile parts are made from medium-carbon steels. These parts include crankshafts, planet pinion shafts, struts, and tie rod ends, among others.

High-Carbon Steel

High-carbon steel contains between 0.60 and 1.0 percent carbon. High-carbon steel is always hardened, but it is also brittle. This steel is often used as cutting tools such as lawn mower blades. High-carbon steels are more costly to make and have poor formability and weldability. Figure 5-6 lists some typical applications of plain carbon steels.

ALLOY STEELS

In the materials field, alloy steels are heat-treatable steels that contain between 1 percent and 4 percent alloying elements. Alloying elements are added to plain-carbon steels for many purposes. Some examples are:

♦ **To improve mechanical properties.** Plain-carbon steels cannot be strengthened beyond about 100,000 psi (6,900 MPa) without significant loss of toughness and ductility. Plain-carbon steels also have poor impact resistance at low temperatures.

♦ **To improve resistance to corrosion and elevated-temperature oxidation.** Plain-carbon steels have poor corrosion resistance for many industrial environments and oxidize readily at high temperatures.

♦ **To improve special properties.** Alloy steels have a higher abrasion resistance and better fatigue behavior than plain-carbon steels.

♦ **To improve hardenability of steels.** Plain carbon steels require high cooling speed for quenching-hardening and are limited to small area sections. Adding alloying elements can increase the depth of hardening and therefore larger sections can be heat treated to reach required hardness and strength.

Figure 5-6.
This table shows some typical applications of plain carbon steel.

TYPICAL APPLICATIONS OF PLAIN CARBON STEEL	
Category	**Applications**
Mild steel	Sheet and strip for presswork; wire and rod for nails and screws; concrete reinforcement bar; steel plate and sections used for structural work.
Medium-carbon steel	Shafts; gears; railroad wheels; suspension and steering parts; intake valves; overrunning clutch cam and hub; transmission kickdown and reverse bands.
High-carbon steel	Forging dies; railroad rails; springs; hammers; saws; cylinder linings; cold chisels; forging die blocks; punchers; shear blades; knives; axes; screwing dies and taps; milling cutters; ball bearings; drills; wood-cutting tools; razors.

Classification of Alloy Steels

In the United States, alloy steels are usually designated by AISI codes similar to the ones used for plain carbon steels. Figure 5-7 lists AISI codes for principal types of standard alloy steels. The first two digits indicate the principal alloying elements or groups of alloying elements. The last two digits show the nominal carbon content of the alloy steel in the same way as plain-carbon steel.

Effect of Alloying Elements

Understanding the effects of adding alloying elements to steel is essential in proper material selection. Generally, adding alloying elements changes the properties of a steel in one of two ways. One change is to directly improve the mechanical properties of the steel, such as hardness, strength, or ductility. The other effect is changing the material's response to heat treatment. This can make the material easier to improve by heat treatment. Heat treatment is covered later in this chapter. Typical applications of selected alloy steels are listed in Figure 5-8.

Manganese increases the strength and hardness of carbon steels. For example, AISI 1330 steel (0.30 percent carbon and 1.75 percent manganese) is used for high-strength bolts.

Silicon strengthens ferrite and therefore increases the strength of the steel. For example, AISI 9260 silicon steel (0.60 percent carbon, 0.88 percent

Figure 5-7.
This table shows type classifications for standard alloy steels.

PRINCIPLE TYPES OF STANDARD ALLOY STEELS	
Type	**Alloy Composition (%)**
13xx	Manganese 1.75
40xx	Molybdenum 0.20 or 0.25; or molybdenum 0.25 and sulfur 0.042
41xx	Chromium 0.50, 0.80, or 0.95, molybdenum 0.12, 0.20, or 0.30
43xx	Nickel 1.83, chromium 0.50 or 0.80, molybdenum 0.25
44xx	Molybdenum 0.53
46xx	Nickel 0.85 or 1.83, molybdenum 0.20 or 0.25
47xx	Nickel 1.05, chromium 0.45, molybdenum 0.20 or 0.35
48xx	Nickel 3.50, molybdenum 0.25
50xx	Chromium 0.40
51xx	Chromium 0.80, 0.88, 0.93, 0.95, or 1.00
51xxx	Chromium 1.03
52xxx	Chromium 1.45
61xx	Chromium 0.60 or 0.95, vanadium 0.13 or min. 0.15
86xx	Nickel 0.55, chromium 0.50, molybdenum 0.20
87xx	Nickel 0.55, chromium 0.50, molybdenum 0.25
88xx	Nickel 0.55, chromium 0.50, molybdenum 0.35
92xx	Silicon 2.00; or silicon 1.40 and chromium 0.70
50Bxx	Chromium 0.28 or 0.50, B denotes boron steel
51Bxx	Chromium 0.80, B denotes boron steel
81Bxx	Nickel 0.30, chromium 0.45, molybdenum 0.12, B denotes boron steel
94Bxx	Nickel 0.45, chromium 0.40, molybdenum 0.12, B denotes boron steel

Figure 5-8.
This table shows some typical
applications of alloy steels.

TYPICAL APPLICATIONS OF ALLOY STEELS		
AISI Code	**Main Chemical Composition**	**Typical Applications**
Manganese steels		
1330	0.30%C, 1.75%Mn	High strength bolts; bicycle frames
Chromium steels		
5130	0.30%C, 0.80%Mn, 0.80%Cr	Steering parts
5169	0.60%, 0.88%Mn, 0.80%Cr	Spring steel
Chromium-molybdenum steels		
4130	0.30%C, 0.50%Mn, 0.95%Cr, 0.20%Mo	Pressure vessels, aircraft structural parts, auto axles, steering knuckles
4140	0.40%C, 0.88%Mn, 0.95%Cr, 0.20%Mo	
Chromium-vanadium steels		
6150	0.50%C, 0.80%Mn, 0.95%Cr, 0.15%V	Valves and springs
Nickel-chromium-molybdenum steels		
8640	0.40%, 0.88%Mn, 0.50%Cr, 0.20%Mo, 0.55%Ni	Auto springs, small machine axles, shafts
Silicon steels		
9260	0.60%, 0.88%Mn, 2.0%Si	Leaf springs

manganese, 2.0 percent silicon) is used to make leaf springs for trucks. A silicon content between 0.5 percent and 5 percent imparts magnetic properties onto steel. A silicon steel of 0.07 percent carbon and 4 percent silicon is used for electrical transformer cores.

Chromium increases strength, fatigue resistance, and hardness. It forms hard and stable carbides. Medium and high-carbon steels with 1.0 percent to 1.50 percent chromium are widely used for gears, axles, shafts, springs, ball bearings, and metal-working rollers.

Nickel has a strengthening effect and is frequently used with chromium. For example, AISI 4340 steel (0.40 percent carbon, 0.70 percent manganese, 0.80 percent chromium, 0.25 percent molybdenum, 1.83 percent nickel) is used for airplane landing gears and heavy truck parts.

There are other carbide forming alloy elements. These include tungsten, vanadium, titanium, and niobium. They precipitate as fine carbide particles, resulting in a significant increase in hardness and strength.

High-Strength Low-Alloy Steels (HSLA Steels)

Automobile manufacturers try to improve gasoline mileage by reducing the total weight of the vehicle. A large percentage of a car's weight is steel. Therefore, efforts have been directed toward producing steel that is stronger and can be used as thinner sections. Conventional alloy-steels are not as weldable and formable as low-carbon steels. Low-carbon sheet steels were developed that were "microalloyed" and rolled under controlled conditions to produce high strength and good weldability. The result is a class of steel known as *high-strength low-alloy (HSLA)* steels.

In general, HSLA steels are not hardened by heat treatment. They are used as supplied. They have low carbon content, less than 0.2 percent. All contain about 1 percent manganese and low concentrations of other elements, such as niobium, titanium, vanadium, and nitrogen. The microalloying elements are used to form precipitates during the hot-rolling process. These fine particles increase material strength, but do not significantly increase manufacturing cost.

Copper is another important element. It strengthens the steel and provides excellent atmospheric corrosion resistance. Copper-bearing HSLA steel is called *weathering steel.*

HSLA steels are primarily intended for structural-type applications where weldability is a major requirement. HSLA steels are used for bridges, towers, railings, and stairs. These steels are marketed not by alloy content, but rather on the minimum mechanical properties. This is because selecting an HSLA steel generally depends on the required mechanical properties.

HIGH-ALLOY STEELS

High-alloy steels are defined as steels having an alloy content of 10 percent or higher. There are two major families of materials in this group, *stainless steels* and tool steels.

Stainless steels were developed to improve corrosion resistance by adding elements such as chromium and nickel. These steels have a desirable microstructure and mechanical properties while they enhance the ability of steels to withstand hostile environments.

A typical cutting tool is subjected to high temperature under normal working conditions. Tool steels resist softening caused by heat. Their high alloy content slows diffusion and coarsening of the hardening carbide.

Stainless Steels

Stainless steels are selected as industrial materials mainly because of their excellent corrosion resistance. This is principally due to high chromium content. Chromium forms a surface oxide film that protects the underlying metal from further corrosion. A small amount of chromium, for example about 5 percent, added to iron provides some corrosion resistance. However, in order to make a steel "stainless," a minimum of 12 percent chromium is required.

The addition of nickel to stainless steel improves its corrosion resistance in neutral or weakly oxidizing media. Nickel in sufficient amounts also improves the ductility and formability by retaining an austenitic structure at room temperature. Molybdenum improves corrosion resistance of stainless steel in the presence of chlorine ions. Aluminum improves high-temperature scaling resistance.

Stainless steel can be divided into three principal groups according to their predominating microstructure. These groups are ferritic, martensitic, and austenitic stainless steels.

Ferritic Stainless Steels

These stainless steels are called ferritic because their microstructure remains mostly as ferrite at normal heat treatment conditions. Ferritic stainless steels are essentially iron-chromium alloys containing 12 to 30 percent chromium and a limited amount of carbon. For example, AISI 430 stainless steel has 17.0 percent chromium and 0.12 percent carbon. These alloys are used mainly as general construction materials where resistance to corrosion and heat is required. Figure 5-9 lists chemical compositions and typical applications of some standard ferritic stainless steels.

Figure 5-9.
This table shows composition and some typical applications of ferritic stainless steels.

CHEMICAL COMPOSITION AND TYPICAL APPLICATIONS OF FERRITIC STAINLESS STEELS						
AISI Code	Chemical Composition wt. %					Typical Applications
	Cr	C(max)	Mo	Al	Other	
405	13	0.08		0.2		(nonhardenable grade) Annealing boxes; quenching racks; oxidation-resistant partitions.
409	11	0.08			Ti6×C	(general-purpose construction stainless) Automotive exhaust systems; transformer and capacitor cases; dry fertilizer spreaders; tanks for agricultural sprays.
430	17	0.12				(general-purpose nonhardenable chromium type) Decorative trim, nitric acid tanks; annealing baskets; heaters; mufflers; range hoods; recuperators; restaurant equipment.
434	17	0.12	1			Modification of type 430 designed to resist atmospheric corrosion in the presence of winter road-conditioning and dust-laying compounds. Used for automotive trim and fasteners.
436	17	0.12	1		Nb5×C	Similar to types 430 and 434. Used where low "roping" or "ridging" required. General corrosion and heat-resistant applications such as automobile trim.
442	20.5	0.2				High chromium steel, principally for parts that must resist high service temperature without scaling. Used for furnace parts; nozzles; combustion chambers.
446	25	0.2				High resistance to corrosion and scaling at high temperatures, especially for intermittent service; often used in sulfur-bearing atmosphere. Annealing boxes; combustion chambers; glass molds; heaters; pyrometer tubes; recuperators; stirring rods; valves.

Ferritic stainless steels provide similar corrosion resistance to the nickel-containing austenitic stainless steels discussed later. However, ferritic stainless steels have low ductility, are sensitive to surface damage, and have poor weldability. These properties limit the use of ferritic stainless steels, especially compared to austenitic stainless steels.

Martensitic Stainless Steels

When a regular steel is cooled fast enough, such as quenched in water, it has a martensitic structure at room temperature. Like cementite (Fe_3C), the martensite structure is very hard. Therefore, the steel is hardened and strengthened. This principle is also applied to improve the mechanical strength of stainless steel. The stainless steel with martensite developed by alloying and heat treatment is called *martensitic stainless steel.*

Martensitic stainless steels are essentially iron-chromium alloys containing 12 to 17 percent chromium. A minimum of 12 percent chromium is required for corrosion resistance. Compared with ferritic stainless steels, martensitic stainless steels contain larger amounts of carbon. This is necessary so that a martensitic structure can be formed after quenching from high temperature. However, the carbon content cannot exceed certain limits. Otherwise, an excessive amount of chromium carbide is formed. This depletes the chromium content of the steel. Depletion of chromium is most detrimental when it occurs on grain boundaries because this tends to cause intergranular corrosion. As a general guideline, when the chromium is 12 percent, the maximum carbon content is 0.15 percent for the best corrosion resistance.

Because of the strengthening effect, martensitic stainless steels are used primarily in applications that require high hardness. These applications include hardware, cutlery, scissors, and surgical tools. Figure 5-10 lists chemical compositions and typical application examples of selected martensitic stainless steels.

Austenitic Stainless Steels

Austenitic stainless steels make up about 65 to 70 percent of the total US stainless steel production. They are essentially iron-chromium-nickel alloys containing 16 to 25 percent chromium and 7 to 20 percent nickel. The most common austenitic stainless steel is type 304. It contains 18 percent chromium and 8 percent nickel and is referred to as 18-8 stainless steel. These stainless steels are called austenitic because their structure remains austenitic at all normal heat treatment temperatures. Some of the nickel can be replaced by manganese and maintain their austenitic structure. A magnet is not attracted to austenitic stainless steel.

Austenitic stainless steels are popular mainly because of their high corrosion resistance and formability. This makes these stainless steels suitable for many industrial applications. Figure 5-11 lists the main chemical compositions and typical applications of some commonly used austenitic stainless steels. Type 302 and 304 are the most widely used stainless steels, in both high-temperature and ambient-temperature applications. Type 316 has higher corrosion resistance and enhanced elevated-temperature strength than 302 and 304. Alloys with increased chromium content, such as type 309, are used primarily for high-temperature applications.

Figure 5-10 .
This table shows composition and some typical applications of wrought martensitic stainless steels.

CHEMICAL COMPOSITION AND TYPICAL APPLICATIONS OF WROUGHT MARTENSITIC STAINLESS STEELS		
AISI Code	**Main Chemical Composition**	**Typical Applications**
403	12.2%Cr, 0.15%C (max.)	Steam turbine blades, jet engine rings, and other highly stressed parts.
410	12.5%, 0.15%C (max.)	(general-purpose heat-treatable type) Machine parts, pump shaft, bolts, bushings, coal chutes, cutlery, fishing tackle, hardware, jet engine parts, mining machinery, rifle barrels, screws, and valves.
420	13%Cr, over 0.15%C	(high-hardenability steel) Springs, tempered rules, machine parts, bolts, mining machinery, scissors, ship belts, spindles, and valve seals.
431	16%Cr, 0.20%C (max.)	Special-purpose hardenable steel used where particularly high mechanical properties are required, such as aircraft fittings, beater bars, paper machinery, and bolts.
440A	17%Cr, 0.72%C	Hardenable to higher hardness than type 420 with good corrosion resistance. Used for cutlery, bearings, and surgical tools.

Figure 5-11.
This table shows composition and some typical applications of wrought austenitic stainless steels.

CHEMICAL COMPOSITION AND TYPICAL APPLICATIONS OF WROUGHT AUSTENITIC STAINLESS STEELS		
AISI Code	**Main Chemical Composition**	**Typical Applications**
301	17%Cr, 7%Ni, 0.15%C (max.)	Used for structural applications where high strength plus high ductility is required, such as railroad cars, trailer bodies, aircraft structures, fasteners, automobile wheel covers and trim, and pole-line hardware.
302	18%Cr, 9%Ni, 0.15%C (max.)	General-purpose austenitic stainless steel used for trim, food-handling equipment, aircraft cowlings, antennas, springs, cookware, building exteriors, tanks, hospital and household appliances, jewelry, oil refining equipment, and signs.
304	19%Cr, 9%Ni, 0.08%C (max.)	Low-carbon modification of type 302 for restriction of carbide precipitation during welding. Used for chemical and food processing equipment, brewing equipment, cryogenic vessels, gutters, downspouts, and flashings.
316	17%Cr, 12%Ni, 0.08%C (max.), 2.5%Mo	Higher corrosion resistance than types 302 and 304 and higher creep strength at high temperature. Used for chemical and pulp handling equipment, brandy vats, fertilizer equipment parts, ketchup cooking kettles, and yeast tubs.
309	23%Cr, 13.5%Ni, 0.20%C (max.)	High-temperature strength and scale resistance. Used for aircraft heaters, heat-treating equipment, annealing covers, furnace parts, heat exchangers, oven linings, and pump parts.

Tool Steels

Although tool steels represent a relatively small percentage of total steel production, they have a unique importance. Tool steels are used in the production and forming of basic materials such as metals, plastics, and wood into desired forms. Some applications of tool steels include drills, deep-drawing dies, shear blades, punches, extrusion dies, rollers, and cutting tools. Some desirable properties of tool steels are high hardness, high wear resistance, good heat resistance, and sufficient strength to work with other materials.

Tool steels have high hardenability. This allows the steel to be cooled slowly, such as air cooled, and still be hardened. Slow cooling protects the steel from cracking and distortion. Tool steels retain their hardness at the high temperatures generated by fast cutting. This is critical for tool applications.

The chemical composition, quenching media, and typical application of selected tool steels are shown in Figure 5-12. Note that the main alloy elements used in tool steels include chromium, molybdenum, tungsten, and vanadium. All the tool steels have high carbon content, except H11 type. Most of the tool steels can be hardened by oil or air.

CAST IRON

Cast iron as an industrial material dates back to at least the fourteenth century. It normally refers to iron-carbon alloys with 2 percent to 4 percent carbon. Cast iron usually contains 1 to 3 percent silicon and other alloying

Figure 5-12.
This table shows composition, quenching media, and some typical applications for tool steels.

CHEMICAL COMPOSITION, QUENCHING MEDIA, AND APPLICATIONS OF TOOL STEELS			
Code	Main Chemical Composition	Quenching Media	Typical Applications
W1	0.6–1.4%C	Water	Tools and die used below 350°F (177°C)
O1	0.9%C, 0.5%Cr, 0.5%W	Oil	Tools and dies requiring less distortion than W1
D1	1.0%C, 12.0%Cr, 1.0%Mo	Air	Wear-resistant, low-distortion tools
H11	0.35%C, 5.0%Cr, 1.5%Mo, 0.4%V	Air	Hot-working dies
M1	0.8%C, 4.0%Cr, 8.0%Mo, 1.5%W, 1.0%V	Air	High-speed tools
T1	0.7%C, 4.0%Cr, 18.0%W, 1.0%V	Air	High-speed tools

elements to control other specific properties. The principal advantages of cast irons are low cost and the ease of casting into complex shapes. Cast irons are very fluid when molten, undergo only slight shrinkage on solidification, and do not form undesirable surface films when poured. They have been widely used as machine bases, engine blocks, engine heads, camshafts, piston rings, manifolds, crankshafts and so on. The principal disadvantages of cast irons are low toughness, low ductility, and poor weldability.

A variety of cast irons can be produced to meet different needs. This is done by varying the balance between carbon, silicon, and other alloying elements and by changing the casting conditions. Five types of cast irons available are white cast iron, gray cast iron, malleable cast iron, ductile cast iron, and high-alloy cast irons. In white cast iron, carbon forms cementite. The other four cast irons can be differentiated from each other by the forms of carbon in the microstructure.

WHITE CAST IRON

White cast iron is produced if the carbon and silicon contents are relatively low and the solidification rate is high. This cast iron shows a white, shining crystalline fracture surface on breaking. This was how the term "white iron" was derived.

Carbon remains combined with iron as iron carbide or cementite. Cementite is hard and brittle and, therefore, makes white cast iron hard and brittle. Because of its high hardness, white cast iron has excellent wear resistance. It is widely used for components such as the grinding balls of a mineral mill, farm equipment, and roller dies.

GRAY CAST IRON

Gray cast iron is formed if the carbon separates during solidification to form graphite flakes. The fractured surface of gray cast iron has a gray color, and therefore the term "gray cast iron." Gray cast irons are the most fluid of the ferrous alloys. As a result, intricate and thin sections can be produced by casting. These irons have lower tensile strength, but relatively high compression strength. They have excellent machinability and are most often used for complex machine bases.

MALLEABLE CAST IRON

Malleable cast iron starts as white cast iron. Then, it is heated to a high temperature and held at the temperature for an extended period of time. The cementite is decomposed into iron and graphite. The graphite precipitates from the solid matrix in the form of irregularly shaped nodules. This process is known as *malleabilizing.* Malleable cast iron is stronger and more ductile than gray cast iron. A downside to malleable cast iron is the high energy cost associated with its production.

A wide range of mechanical properties can be obtained in malleable iron by varying the malleabilizing heat treatment. This type of cast iron is especially suitable for small, thin-sectioned parts. A typical application is automotive differential housings because they require high strength and impact resistance.

DUCTILE CAST IRON

Ductile cast iron has nodular graphite that is formed on solidification. Many manufacturers have switched to this cast iron because of high energy cost of producing malleable cast iron. Adding a very small amount of magnesium or rare earth to molten iron results in spheroidal graphite in the cast iron.

Because of the shape of the graphite, ductile cast iron is sometimes called *nodular cast iron* in the United States and *spherulitic graphite iron* in England. The spherical nodules of graphite make the iron much tougher and more ductile than gray cast iron of similar chemical composition.

Ductile cast iron is replacing steel in many applications because of its low production cost and good castability. Ductile cast iron is used for many components where high strength and toughness are needed, such as automobile engine crankshafts.

HIGH-ALLOY CAST IRON

This group of cast irons includes the highly-alloyed white cast irons, modified gray cast irons, and ductile cast iron. Large amounts of alloying elements are added to achieve certain special performance, such as high abrasive resistance, heat resistance, and corrosion resistance. For example, 30 percent chromium in cast iron produces a highly wear- and corrosion-resistant alloy. This is used in mining and agricultural applications where abrasive environments exist. Cast iron with a high silicon content produces a corrosion-resistant cast iron used for acid containers and chemical pumps.

HEAT TREATMENT

Heat treatment is a process of heating and cooling certain alloys to obtain desired properties. A blacksmith dips the red hot metal being worked on into water to harden the metal. Similarly, modern foundries heat treat alloys to change their properties. Nearly every steel product is heat treated before final finishing. Products requiring high strength and hardness, such as gears, chisels, plows, ball bearings, and so on, are all heat treated.

When AISI 1080 steel is slowly cooled from 1550°F (845°C), the steel has a tensile strength of 112,000 pound per square inch (psi) and a hardness of 192 HB. However, quickly cooling the metal from the same temperature by water-quenching increases the strength of the steel to more than 200,000 psi and a hardness to 680 HB.

Heat treatment is divided into groups according to how a metal is heated and cooled. Different heating and cooling methods result in different microstructure and, therefore, different material properties. The heat treatment groups are quenching, tempering, austempering, normalizing, annealing, and surface hardening.

QUENCHING

Quenching involves heating a steel to above the critical temperature, holding it there for a period of time, and rapidly cooling the metal. The heating process is called austenitizing. The microstructure that results from quenching a steel is *martensite.* Martensite is very hard, which increases the strength and hardness of the steel. However, quenching decreases the toughness and ductility of the steel. There are three questions that need to be answered to effectively quench a steel.

The first question is what critical temperature is required. For the best strengthening effect, an austenite structure should be achieved before cooling. However, high temperature also results in a fast crystal grain growth in the alloy. Large grain sizes decrease the effectiveness of quench hardening and deteriorate the material properties. Thus, the steel should be heated high enough to produce austenite, but not high enough to produce unnecessary grain growth.

The second question that needs to be answered is how long the material should be held at the austenitizing temperature before quenching. The purpose of keeping the part at the austenitizing temperature for sufficient time is to produce uniform microstructure and chemical element distribution within the material. In practice, the holding time is determined by the thickness of the metal. Generally, one hour per inch of the cross section is the rule of thumb. The minimum holding time is usually 30 minutes.

The third question that needs to be answered is what kind of quenching medium is best for the application. Different media cool at different rates. Common quenching media are brine, water, polymer, oil, and air. Of these, brine cools the fastest and air the slowest.

Water as a quenching medium is inexpensive, readily available, and is easy to dispose. It provides a rapid quench, but can cause complex parts to distort or crack. Therefore water is generally used to quench symmetrical parts with low hardenability, such as plain carbon steel bars.

Brine is a water solution containing a percentage of salt. A common brine solution used for quenching contains 10 percent sodium chlorides. Brine cools the piece even faster than water. Thus, cracking and distortions are more likely to occur. Brine is also corrosive. This means that equipment must be protected from corrosion, which can be expensive.

Mineral oils are used for oil quenching. Oil cools much slower than water. This reduces the dangers of cracking and distortion associated with water and brine quenching. Oil quenching is commonly used for complex parts made of low-alloy steels.

Polymer quenching uses solutions of water and organic polymers, such as polyalkene glycol. Polymer quenching rates are between those of water and oil quenching.

Air quenching uses a fast-moving stream of air. It is the slowest method and can only be used for tool steels with a large percentage of alloying elements.

TEMPERING

Quenching hardens a steel, but also makes it brittle. Internal stresses are also created as a result of the fast cooling. Thus, there is a good chance of further distortion and cracking. This problem of cracking and distortion can be solved by tempering. *Tempering* is when a quenched part is reheated to a temperature much lower than quenching temperature, held there for a long period, and then cooled.

Tempering must be done as soon as possible after quenching to reduce the risk of cracking. Parts are usually removed from the quenching medium when they are warm to the touch and then tempered immediately.

The holding time at the tempering temperature is one hour per inch of cross section. Most steels are tempered between 400°F (205°C) and 1100°F (595°C). The microstructure attained after quenching and tempering is tempered martensite. This is more ductile than fresh martensite.

AUSTEMPERING

Most hardenable steels need to be quenched and tempered to achieve a desired balance of hardness and ductility. *Austempering* is a process that combines quenching and tempering. It involves quenching to a temperature, holding the part at this temperature until transformation is completed, and then air cooling the part. The resulting microstructure is known as *bainite.* It has similar mechanical properties to tempered martensite.

Austempering is used to reduce quenching distortion and cracking while creating a tough, strong steel. It is especially suitable for heat treating thin sections of plain-carbon and low-alloy steels.

NORMALIZING

Normalizing consists of heating the steel to a completely austenitic structure followed by air cooling. The heating temperature and holding time are the same as austenitizing before quenching. Normalizing yields smaller grains and finer pearlite than slow-cooling in a furnace. In general, the pearlite is softer than martensite and bainite. Therefore, normalized steel is not as hard and strong as quench-and-tempered or austempered steels.

ANNEALING

There are some cases that require materials with improved ductility, no internal stress, and good machinability and formability. This can be achieved with annealing. *Annealing* consists of heating a steel or cast iron to a high temperature and then cooling it slow enough to prevent the formation of a hardened microstructure. After annealing, the material is much softer than the quenched steel. Usually, the cooling is achieved by turning the furnace off and letting the furnace cool down to ambient temperature with the door closed. For example, an annealing process recommended for AISI 1040 is to heat the steel to 1550°F (845°C) and furnace-cool it at a rate not exceeding 50°F (30°C) per hour.

EFFECTS OF ALLOYING ELEMENTS ON HARDENABILITY

The *hardenability* of a steel is how easy it is to achieve a martensite structure in the steel. The better the hardenability, the deeper the hardening that can be achieved. Also, a material with higher hardenability can be cooled more slowly to develop the same strength and hardness. Slower quench rates greatly decrease the chance of distortion and cracking on the products. Adding alloying elements to a steel significantly improves hardenability. For example, the alloyed steel type T1, a tool steel, can be hardened by air cooling.

All common alloy elements, except cobalt, increase the hardenability of steel. See Figure 5-13. Manganese is the most effective alloy to improve hardenability, followed by molybdenum, chromium, silicon, and nickel.

SURFACE HARDENING

Hardness of a steel must be increased if high wear resistance is required, such as knifes and gears. However, increasing hardness of the steel by quenching results in decreased ductility. The product may be too brittle for a satisfactory service life. To solve this dilemma, a hardening technique can be used to produce only a hard surface layer. The internal metal retains the original toughness and ductility. The key to success of surface hardening is heating the surface layer quick enough so that the interior of the part is not affected.

Surface hardening can be done in two ways. One is to change the microstructure of the surface layer by localized heating on the surface and then quenching. The other way is to introduce carbon and alloy elements to the surface layer by diffusion. Some surface hardening techniques combine the two ways.

Flame Hardening

The principle of *flame hardening* is to rapidly heat the surface of the steel with a direct flame until austenitic is formed and then quickly quench the piece so that its surface is transformed to a hard martensitic phase. The heating and quenching must be done very rapidly. If enough heat is applied to the surface to reach the austenitizing temperature, a hard martensite outer shell is formed on quenching. Suitable materials for flame hardening must have sufficient carbon content, usually 0.40 percent, to allow hardening.

Figure 5-13.
This graph shows the relative effectiveness of improving hardenability of steel by various alloy elements.

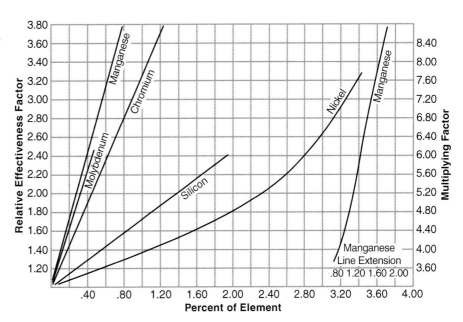

Flame hardening is applied to a wide variety of ferrous alloys. Some examples are:

♦ Parts so large that conventional furnace heating is impractical, such as large gears, dies, and rolls.

♦ Small sections or areas of a part that require surface hardening, such as the ends of valve stems, push rods, and the wear surfaces of cams and levers.

Principal advantages of flame hardening are increased wear resistance and improved fatigue strength. Distortion of the part is a potential problem.

Induction Hardening

Induction hardening is done in the same way as flame hardening, except the heat source is a magnetic field. The magnetic field is generated by high-frequency alternating current. The rapidly alternating magnetic field induces current flow within the steel surface. The current is then converted to heat energy on the surface layer. As a result, the surface layer is rapidly heated to austenitizing temperature. Subsequent quenching results in a hardened surface layer.

The heating rate by induction heating is faster and more uniform than flame. Therefore, it results in less distortion. However, the induction heating coil has to be designed to fit the shape of the part to be the most effective. Induction hardening is used when many symmetrically shaped parts are to be surface-hardened. Examples include crankshafts, camshafts, axle shafts, gears, cams, and valve seats.

Laser Hardening

A laser is an ideal power source for surface hardening. It quickly generates an intense, localized heat. In *laser hardening,* a laser beam travels at a controlled speed across the surface to be hardened. The heating is so localized that the bulk of the material acts as a heat sink to cool the part. No additional quenching is required. As a result, laser hardening produces a shallow layer of martensite. The hardened layer is usually thinner than that produced by flame or induction hardening. An advantage of laser hardening is the ability to harden relatively small areas in a complex part, such as bores in an engine block. A disadvantage is the laser and related equipment can be expensive.

Carburizing

The *carburizing* process starts by adding carbon into the surface layer through diffusion. The carbon content of the surface layer is increased to about 0.8 to 1.0 percent. Typical carbon content of steel for this purpose is 0.10 to 0.2 percent. Carbon is introduced into the steel at high temperature, where the steel is austenitic. The temperature range is usually 1550°F to 1750°F (850°C to 950°C). After carburizing, the work-piece is usually quenched to produce a steel with a hard martensitic surface layer. Because the carbon content for the bulk material is low, only the surface layer with high carbon content is hardened. Carburizing is widely used in industry for machine components where a contact surface must resist wear and fatigue.

Examples include gears, bearings, and shafts. Various processes have been developed to introduce carbon into the surface layer. These include pack carburizing, liquid carburizing and gas-carburizing.

The carburizing process depends on the movement of carbon atoms from the surrounding medium to the surface of the steel. In *pack carburizing,* the part is placed into a steel container. The container is then packed with granules of activated charcoal. This completely surrounds the part with charcoal for the diffusion process. *Liquid carburizing* uses a melted salt to carry carbon into the surface layer. This process is also known as *salt carburizing.* In *gas carburizing,* carbon is introduced through gaseous chemical reaction. Gas carburizing is the fastest way to add carbon to a surface layer. Pack carburizing is the slowest way to add carbon.

For carburizing, a part needs to be made from a special carburizing steel in which some alloying elements are added. With the alloy elements, the grain growth in the steel at high temperatures is inhibited. Therefore, the best surface hardening results can be achieved while maintaining the strength and ductility of original material.

Nitriding

Nitriding is one of the most useful surface-hardening techniques. In this process, nitrogen is diffused into the surface of the steel being treated. The reaction of the nitrogen with the steel forms a very hard iron and nitrogen compound alloy. The resulting nitride surface layer can be harder than even the hardest tool or carburized steels. Two outstanding advantages of this process over all other hardening processes is that much lower temperatures are used and the hardness is achieved without any quenching. An additional advantage is the nitrogen atmosphere used in the hardening process prevents oxidation and scaling.

SUMMARY

Ferrous alloys consist of the largest portion of industrial materials in use today. They are the most economical materials available for construction, heavy machinery, and other parts requiring load-bearing capacities. Ferrous alloys include different types of steels and cast irons. Each offers a broad range of properties to meet various industrial needs, including mechanical strength, ductility, corrosion resistance, and wear resistance. Properties of steel and cast iron can be altered by various alloying elements and heat treatments. The relationship between composition, structure, and properties is the key to proper material application.

IMPORTANT TERMS

δ Iron
α Iron
γ Iron
Alloy Steel
Alloying Elements
Annealing
Austempering
Austenite
Austenitic Stainless Steels
Austenitizing
Bainite
Carburizing
Cast Iron
Cementite
Coke
Ductile Cast Iron
Ferrite
Ferrous Alloys
Flame Hardening
Gas Carburizing
Gray Cast Iron
Hardenability
Heat treatment
High-Alloy Steel
High-Carbon Steel
High-Strength Low-Alloy (HSLA)
Impurities
Induction Hardening

Iron Ore
Iron-Iron Carbide Phase
   Diagram
Laser Hardening
Liquid Carburizing
Low-Carbon Steel
Malleabilizing
Malleable Cast Iron
Martensite
Martensitic Stainless Steel
Medium-Carbon Steel
Nitriding
Nodular Cast Iron
Normalizing
Pack Carburizing
Pig Iron
Plain Carbon Steel
Quenching
Salt Carburizing
Spherulitic Graphite Iron
Stainless Steel
Steel
Superalloys
Tempering
Tool Steel
Weathering Steel
White Cast Iron

QUESTIONS FOR REVIEW AND DISCUSSION

1. What are the differences between steel and cast iron?
2. List three reasons why iron and steel are widely used in industrial applications.
3. Define plain carbon steel, alloy steel, and high-alloy steel.
4. Why does pig iron have a high carbon content?
5. Define ferrite and austenite.
6. What is the chemical composition of cementite?
7. Is cementite harder or softer than ferrite?
8. What is the carbon content of AISI 1040 steel?
9. What is the carbon content of AISI 4140 steel?
10. Which is a plain carbon steel, AISI 1040 or AISI 5140?
11. What is stainless steel used for?
12. Why are there different kinds of stainless steels?
13. What type of stainless steel would be a good choice for a hunter's knife?
14. If a magnet is not attracted to a stainless steel, what type is it? If the magnet is attracted to the stainless steel, what type is it?
15. Why does tool steel need to be stable at high temperatures?
16. How do the structures and mechanical properties of white cast iron and gray cast iron differ?
17. Why is the mechanical property of a cast iron improved when the graphite morphology is changed from flake to a nodular shape?
18. What are the main reasons for adding alloying elements to steel?
19. If you have a round bar 2" in diameter with a length of 4", how long should the holding time be before quenching?
20. Using the phase diagram, determine the austenitizing temperature for AISI 1045.
21. Define austempering. What are the advantages of using austempering instead of quenching and tempering?
22. If you find a steel piece is hard to machine, what can you do to improve its machinability?
23. Suppose you have a very large gear that will not be mass produced. The gear requires high wear resistance. What kind of hardening technique would be best? Why?
24. Define carburizing. Why can carburizing harden the surface, but not the interior of a part?
25. Why is nitriding better than carburizing for many applications?

FURTHER READINGS

1. Smith, W.F. Structures and Properties of Engineering Alloys (2nd ed.). New York: McGraw-Hill, Inc. (2000).

2. Smith, W.F. Foundations of Materials Science and Engineering (2nd ed.). New York: McGraw-Hill, Inc. (1992).

3. Flinn, R.A. and Trojan P.K. Engineering Materials and Their Applications (4th ed.). Boston: Houghton Mifflin Co. (1994).

INTERNET RESOURCES

http://www.steel.org
American Iron and Steel Institute

http://www.asm-intl.org
American Society for Metals, International

http://www.tms.org
The Minerals, Metals & Materials Society

http://www.astm.org
American Society for Testing and Materials

http://www.bethsteel.com
Bethlehem Steel

Chapter 6

Nonferrous Metals and Alloys

KEY CONCEPTS

Upon completion of this chapter, you should understand:
- ➤ Basic characteristics of copper, aluminum, titanium, zinc, magnesium, and alloys.
- ➤ Classification and typical applications of copper and copper alloys.
- ➤ Classification, heat treatment, and applications of aluminum alloys.
- ➤ Titanium alloys and their applications.
- ➤ Advantages and disadvantages of using zinc or magnesium alloys.
- ➤ Basic applications of precious metals.

Nonferrous metals and alloys are all metals and alloys other than ferrous metals and alloys. Ferrous metals and alloys are discussed in Chapter 5. Nonferrous metals and alloys comprise three-fourths of the known elements in nature. While ferrous alloys (iron and steel) are the most often used metals, nonferrous metals and alloys are receiving more and more attention because of their unique properties. For example, aluminum alloys make lightweight airplane structures possible. Copper is a good conductor and makes the transmission of electricity to millions of homes possible. This chapter discusses widely used nonferrous metals and alloys, including copper, aluminum, titanium, zinc, manganese, and precious metals.

COPPER AND COPPER-BASED ALLOYS

Historically, copper was one of the first industrial metals used by humans. Unlike most other metals, copper exists in nature in its metallic form. Large amounts of copper could be obtained by simply crushing rocks and removing chunks of solid copper. However, relatively little of this form of copper is left today. Most of today's supply of copper comes from copper ores that may contain only 5% copper. The cost of copper is increasing. It is currently about five times the cost of low-carbon steel. Nonetheless, because of their unique properties, copper and copper alloys are indispensable industrial materials.

Copper is widely used in both its pure condition and in alloys. Figure 6-1 illustrates some examples of various products made of pure copper or copper alloys. Pure copper has a superior combination of electrical conductivity, corrosion resistance, ease of fabrication, and reasonable strength. The wide variety of brasses and bronzes offer various combinations of strength, ductility, fabrication characteristics, corrosion resistance, and wear resistance. In this section, the classification, properties, and applications of copper and copper alloys are discussed.

CLASSIFICATION

Copper is commonly used as pure copper and as copper alloys. Brass, bronze, and beryllium copper are the major copper alloys. Copper and copper alloys are classified according to a system administered by the Copper Development Association (CDA). All copper and alloys are divided according to their chemical compositions, as shown in Figure 6-2. In this system,

Figure 6-1.
Some examples of various products made of pure copper or copper alloys.
(Courtesy of Brush-Wellman)

CLASSIFICATION OF COPPER AND COPPER ALLOYS	
Code	Description
Wrought copper and copper alloys	
C1xx00	Pure coppers and high-copper alloys having more than 96 percent copper but not fitting into any alloy category, including beryllium-copper
C2xx00	Copper-zinc alloys (brasses)
C3xx00	Copper-zinc-lead alloys (leaded brasses)
C4xx00	Copper-zinc-tin alloys (tin brasses)
C5xx00	Copper-tin alloys (phosphor bronzes)
C6xx00	Copper-aluminum alloys (aluminum bronzes), copper-silicon alloys (silicon bronzes), and miscellaneous copper-zinc alloys
C7xx00	Copper-nickel and copper-nickel-zinc alloys (nickel silvers)
Cast copper and copper alloys	
C8xx00	Cast copper, cast high-copper alloys, the cast brasses of various types, cast manganese-bronze alloys, and cast copper-zinc-silicon alloys
C8xx00	Cast copper-tin alloys, copper-tin-lead alloys, copper-tin-nickel alloys, copper-aluminum-iron alloys, and copper-nickel-iron and copper-nickel-zinc alloys

numbers from C10000 through C79900 designate wrought alloys. Numbers from C80000 to C99900 designate cast alloys. The last two digits are always zero. The zeros were added by the CDA in late 1970s to conform with Unified Numbering System (UNS). Unlike AISI steel codes, the alloy designation number for copper does not indicate any properties or composition of the alloy.

Pure Copper

Copper (Cu) as an element has a face-centered cubic (FCC) crystal structure. It has excellent conductivity, second only to silver among metallic elements. Other desirable properties of pure copper are its high corrosion resistance, ease of fabrication, reasonable tensile strength, and good soldering and joining characteristics. During the refining process, oxygen and other impurities may be absorbed into the metal. Industrial coppers can be divided into three major categories according to their oxygen and impurity contents. These categories are discussed in the next sections.

Electrolytic Tough-Pitch (ETP) Copper

Electrolytic tough-pitch (ETP) copper has a minimum of 99.9 percent copper and an average of 0.04 percent oxygen. The normal limits of oxygen in ETP copper are between 0.02 and 0.05 percent. ETP is the least expensive of the industrial coppers. It is used extensively for production of wires, rods, plates, and strips.

Oxygen-Free Copper

Oxygen-free copper is produced in an environment that prevents oxygen from entering the copper. This environment is usually an atmosphere of carbon monoxide and nitrogen. As a result, oxygen-free copper contains 99.95 percent copper. Due to the special processing, oxygen-free copper is more expensive than ETP copper.

The electrical conductivity of oxygen-free copper is about the same as electrolytic tough-pitch copper. Eliminating Cu_2O in the copper increases conductivity. However, this is offset by an increase of other impurities, such as iron.

Deoxidized Copper

With the addition of sufficient phosphorous, all the available oxygen in the copper is converted to phosphorus pentoxide (P_2O_5). This produces *deoxidized copper*. Deoxidized copper has high electrical conductivity. This is because the phosphorus residue present is normally less 0.009 percent.

Brass

Brass is an alloy of copper with up to 40 percent zinc. Brass has a distinctive yellow color. Variations in composition result in desired color, strength, ductility, machinability, and corrosion resistance. The most common brass is 60 percent copper and 40 percent zinc. This is called **mains metal.** Brass with 30 percent zinc is called **yellow brass.** Yellow brass is often alloyed with lead to improve machinability. Those brasses are called **lead brasses.** **Red brass** contains about 15 percent zinc and is more copper colored than yellow. In the cast form, red brass usually is an alloy of 85 percent copper, 5 percent zinc, 5 percent tin, and 5 percent lead. Many brasses contain other alloying elements such as aluminum, tin, nickel, and arsenic for special purposes.

Bronze

Bronze was originally a copper-tin alloy. Now, however, copper can be alloyed with almost any alloying element except zinc to create bronze. Modern bronze can more properly be called **tin-bronze.** The tin-bronzes are also commercially called **phosphorus bronzes** because they frequently contain about 0.3 percent phosphorus as a result of the refining process. Tin-bronze has desirable properties such as high strength, wear resistance, and good resistance to seawater corrosion.

Silicon bronze contains 1 to 3 percent silicon in place of tin. It is an important alloy for maritime applications and high-strength fasteners.

Aluminum bronze contains up to 10 percent aluminum in place of tin. These alloys are quite hard and tough. Aluminum bronze generally has a high tensile strength as well. Iron is sometimes added to aluminum bronze to increase the strength and hardness.

Beryllium Copper

The addition of small amounts of beryllium to copper creates a family of high strength copper alloys. *Beryllium copper* alloys can be heat-treated to produce a tensile strength as high as 212,000 psi (1462MPa). This is the highest strength developed in commercial copper alloys. For comparison, a quenched and tempered AISI 1040 steel has a tensile strength of 130 ksi. Commercial beryllium copper alloys contain between 0.6 and 2 percent beryllium with 0.2 to 2.5 percent cobalt.

PROPERTIES AND APPLICATIONS

The different copper alloys have different properties. These properties make them ideal for different types of applications. Properties and applications of different coppers and alloys are covered in the next sections.

Pure Copper

Pure copper is an ideal candidate for electrical or thermal conductors. It is estimated that about 50 percent of all copper used in the United States is for electrical conductors. Typical products include bus bars, wires, contacts, switches, printed circuit boards, and transistor bases. Therefore, electrical conductivity is usually the most important property. To keep conductivity high, impurities, including oxygen, must be kept low.

For some applications with possible temperature increases, small amounts of other elements, such as silver and cadmium, may be added to prevent annealing. Pure copper has found many applications for heat transfer devices, such as automotive radiator cooling fins, air conditioners, gas and heater liners, and gaskets.

Pure copper is easy to deform and bend. Some plumbing parts, such as tubes and fittings, are made of pure copper because of its ease of deformation. Pure copper also has high corrosion resistance. This is also a desirable property for plumbing parts.

Brass

The strength and hardness of brasses increase with the amount of zinc alloyed with the copper. The best combination of ductility and strength occurs at 70 percent copper and 30 percent zinc. This alloy can be used for deep drawing and is called *cartridge brass.* It is also widely used for other applications, such as radiator cores and lamp fixtures.

Other alloying elements may be added to improve properties. For example, addition of manganese, iron, and tin to copper increases the tensile strength of brass. The addition of 3 percent of lead improves the machinability of brass. However, the ductility of the brass is reduced due to the addition of lead.

Bronze

Tin-bronze has excellent cold workability and a good combination of strength, ductility, and corrosion resistance. It is used in electrical contacts, flexible hoses, pole-line hardware, chemical hardware, and welding rods.

Aluminum bronzes are quite hard and have high tensile strengths. They resist wear and fatigue and have excellent corrosion resistance.

Silicon bronzes have properties similar to aluminum bronze. For many applications where resistance to seawater corrosion is needed, silicon bronzes are low cost substitutes for tin-bronzes.

Beryllium Copper

Beryllium copper offers the highest strength among all copper alloys. Since the amount of beryllium added is relatively low, the conductivity of beryllium copper is higher than most other copper alloys. Beryllium coppers are suitable for electrical contact springs that require both high electrical conductivity and mechanical strength. Beryllium copper is also used for tools that must be hard but cannot spark, such as those used in the chemical industry. The corrosion and fatigue resistance of beryllium coppers makes them useful for springs, gears, diaphragms, and valves.

However, there are two major disadvantages of using beryllium copper. Though these alloys contain only a small amount of beryllium, they can be expensive. Another consideration, especially for casting, is that beryllium vapor is toxic. Special safety protection is required.

ALUMINUM AND ALUMINUM ALLOYS

Aluminum is the third most commonly used metal, behind iron and steel. Aluminum has a unique combination of properties that make it one of the most versatile industrial materials. Aluminum is lightweight, yet some of its alloys are stronger than structural steel. For example, many airplane designs rely on high-strength, lightweight aluminum alloys. Aluminum has good electrical and thermal conductivity and high reflectivity to both heat and light. It is highly corrosion-resistant under many service conditions and it is nontoxic. Aluminum can be cast and formed into almost any shape and can provide a variety of surface finishes. Figure 6-3 shows typical applcations of aluminum alloys in an automobile body.

The next sections discuss the classification, properties, and applications of aluminum alloys. Heat treatment is an important way to strengthen some aluminum alloys. Thus, heat treating aluminum is also discussed in the next sections.

CLASSIFICATION

A four-digit designation identifies wrought aluminum and wrought aluminum alloys, as in Figure 6-4. The first digit indicates the group of the aluminum alloy according to the principal alloying elements added. For example, the first 6 in the 6061 designation indicates an aluminum-magnesium-silicon alloy.

Figure 6-3.
Typical applications of aluminum alloys in an automobile body. All parts shown here are aluminum. (The Aluminum Association, Inc.)

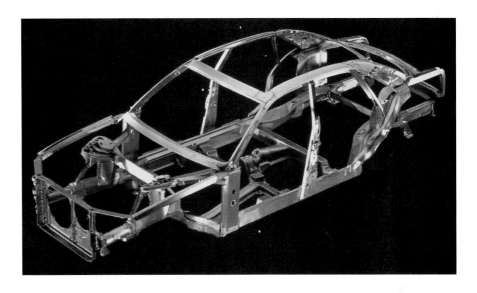

Figure 6-4.
This table shows the groups of aluminum alloys.

WROUGHT ALUMINUM ALLOY GROUPS	
Code	**Description**
1xxx	Commercially pure aluminum (99% min.)
2xxx	Copper (major alloying element)
3xxx	Manganese
4xxx	Silicon
5xxx	Magnesium
6xxx	Magnesium and silicon
7xxx	Zinc
8xxx	Other elements
9xxx	Unused series

For wrought aluminum alloys, a letter code is normally added after the four-digit designation to indicate the treatment, or *temper*, used on the aluminum alloy. For example, the T6 in the 6061-T6 designation indicates the aluminum-magnesium-silicon alloy was solution treated and furnace age hardened. Pay special attention to the heat treatment history of an aluminum alloy. This determines mechanical properties of heat-treatable aluminum alloys. Details of the treatment designation can be found in the aluminum handbook.

A similar system is used to designate cast aluminum alloys, as shown in Figure 6-5. However, the numbers may represent different designations. The first digit indicates the alloy group. The second and third digits identify an alloy within the group. The last digit, separated by a decimal, shows the product form.

PROPERTIES AND APPLICATIONS

The tensile properties of aluminum alloys vary from a tensile strength of 13 ksi (90 MPa) for pure aluminum (designation 1100) to as high as 98 ksi (676 MPa) for an age-hardened aluminum-zinc-copper-magnesium alloy (designation 7001-T6). Pure aluminum is used where high conductivity is required, but strength is not critical, such as wires and bus bars.

The 3xxx and 5xxx series have tensile strength from 16 ksi (110 MPa) to around 50 ksi (345 MPa). For example, 3003 alloy is a general-purpose alloy that has good formability, weldability, and corrosion resistance. It is widely used for sheet-metal parts such as covers, chutes, guards, wire ways, and switch boxes. If these properties are desired but a higher strength is needed, the 5xxx series is a good choice. For example, 5052 alloy is widely used for storage tanks and structural members.

The 6xxx series is considered a medium-strength alloy. The 2xxx and 7xxx series are high-strength alloys. For these alloys, high strength means poor formability and fabricability. Aluminum alloys such as 6061 and 6063 have good weldability and moderate formability. However, high-strength aluminum alloys such as 2024 and 7075 have lower weldability and poor formability. Aluminum alloys 6061 and 6063 are normally used for structural components unless the strength of the 2xxx and 7xxx series alloys is absolutely necessary. Examples of parts made from 2xxx, 6xxx, and 7xxx alloys include machine frames, truck bodies, and architectural trim extrusions.

Figure 6-5.
This table shows cast aluminum alloy designations.

CAST ALUMINUM ALLOY DESIGNATIONS	
Code	**Description**
1xx.x	Aluminum, 99.5% minimum
2xx.x	Copper
3xx.x	Silicon, with added copper and / or magnesium
4xx.x	Silicon
5xx.x	Magnesium
6xx.x	Unused series
7xx.x	Zinc
8xx.x	Tin
9xx.x	Other elements

Another consideration in selecting aluminum alloys for industrial applications is the modulus of elasticity. Aluminum alloys are not as stiff as steel and copper. The elastic deflection for aluminum is three times as much as steel of equal size. All aluminum alloys have low modulus of elasticity because the modulus of elasticity is not sensitive to the composition of an alloy.

A surface oxide film gives aluminum its corrosion resistance. For indoor applications, an aluminum alloy can remain corrosion-free for an indefinite period or time. Many household appliances have aluminum trim, handles, housings, or nameplates. Corrosion resistance in outdoor applications depends on chemical and moisture conditions. Usually, corrosion rates in outdoor applications are high for the first two years or so until the aluminum surface is uniformly covered by an oxide film. All aluminum alloys are attacked by immersion in seawater or saltwater sprays.

Aluminum alloys are not widely used in plumbing systems. Water carried in pipes can contain heavy metal ions, such as copper, tin and nickel. These ions cause pitting. *Pitting* is when corrosion occurs and penetrates into the material through a localized hole.

Many aluminum alloys are surface-coated for improved corrosion resistance. Moreover, surface coating makes aluminum alloys suitable for decorative applications especially in construction. One of the most common coating practices is anodizing. *Anodizing* is when a thick oxide coating is formed on the surface by means of an electrochemical process. Usually, an anodized coating is porous. Immersing the anodized coating in hot water converts the oxide to a hydrated form that swells and reduces the pore size. If a dye is put in the sealing water, the coating takes on that color. Architectural shapes and automotive trims are often colored by this process.

HEAT TREATMENT

There are two types of aluminum alloy based on response to heat treatment. The non-heat-treatable grade cannot be hardened using heat treating. This group includes pure aluminum (designation 1*xxx*), aluminum-manganese alloys (designation 3*xxx*), and aluminum-magnesium alloys (designation 5*xxx*). The heat-treatable aluminum alloys can be hardened using heat treating. This group includes aluminum-copper alloys (designation 2*xxx*), aluminum-silicon alloys (designation 4*xxx*), aluminum-magnesium-silicon alloys (designation 6*xxx*), and aluminum-zinc (designation 7*xxx*) alloys.

Solution Treatment and Age Hardening

Solution treatment is when an alloyed aluminum is heated to dissolve the alloying elements into the aluminum and then quenched. For example, when the alloy of aluminum with 4 percent copper is heated to 930°F (500°C), the copper dissolves into the aluminum. When the aluminum-copper alloy is quenched in water, the copper is locked in the aluminum matrix. Solution treatment does not increase the hardness of the aluminum alloy because there is no hardening phase formed upon quenching. This is totally different from the quenching treatment of steels.

However, if the aluminum alloy is held at a low temperature, such as 400°F (204°C), for several hours, the copper precipitates from the solid solution to form $CuAl_2$. The intermetallic compound $CuAl_2$ is a hardening phase, which strengthens the aluminum alloy. This process is known as *age hardening.* The final hardness and strength of the aluminum alloy are dependent on the size and distribution of the precipitated phase. The age hardening process needs to be controlled in terms of temperature and time so that the optimum property can be achieved.

Annealing

Annealing is performed on cold-worked alloys to aid forming. Deep drawing involves a large deformation on an aluminum alloy. It causes hardening on the material. Sheet-metal parts subject to deep drawing operations may require annealing between drawing steps. Otherwise, the material may crack at the strain-hardened locations. Moreover, castings with varying cross sections may solidify with different structures. Annealing gives the casting the same structure throughout and makes the mechanical properties more uniform.

Annealing is done by heating the aluminum alloy to a high temperature, followed by slow cooling. This is similar to the annealing process of steels. However, different aluminum alloys require different annealing temperatures.

TITANIUM ALLOYS

Titanium and its alloys are relatively new industrial materials. They have been in use as structural materials only since the early 1950s. Titanium was discovered in 1791, but it was not produced in metallic form until 1910. It remained a laboratory curiosity until commercial processes were developed for its manufacture in the 1940s. Since then, the production processes have been improved significantly and costs lowered.

Titanium alloys are attractive because they have a high strength-to-weight ratio. Titanium weighs only a little over half of a steel. Its density is 0.16 lb/in^3 (4.5 g/cm^3) whereas iron is 0.28 lb/in^3 (7.87 g/cm^3). The mechanical properties of titanium can be better than many alloy steels. Thus, structures can be made much lighter while supporting the same level of stress or load. Moreover, titanium has a much higher modulus of elasticity than other light metals, such as magnesium and aluminum. This property is very useful for structures requiring rigid materials. Titanium also has excellent high-temperature properties to about 1020°F (550°C), and excellent corrosion resistance, including oxidizing acids and chloride media.

Titanium can be used as pure metal or in alloy form. Commercially pure titanium is 99.0 percent to 99.5 percent titanium. Oxygen is present in the metal for its strengthening effect. The level of oxygen can be adjusted to modify the strength of the titanium. Commercially pure titanium has lower strength, higher corrosion-resistance, and is less expensive than titanium alloys. It is used primarily when strength is not the main requirement. Some examples are airframes, chemical plant parts, marine parts, and surgical implants.

About 70 percent of all titanium applications use the Ti-6%Al-4%V alloy. This alloy contains 90% titanium, 6% aluminum, and 4% vanadium. It can be readily welded, forged, and machined. It is available in a wide variety of mill product forms, such as sheets, extrusions, wires, and rods. This alloy is heat-treatable to achieve an ultimate tensile strength of 165 ksi and has good metallurgical stability up to 900°F (480°C). It is typically used as rocket motor cases, blades and disks for aircraft turbines, pressure vessels, gas and chemical pumps, cryogenic parts, marine components, and steam turbine blades. Alloying elements such as tin, iron, and copper may sometimes be added to increase the strength.

ZINC AND MAGNESIUM ALLOYS

Zinc and magnesium are metals with low melting temperatures. Because of their low melting point, they are easy to cast into complex shapes efficiently. A molten metal flows more easily if it is heated high above its melting point. Since zinc and magnesium have low melting points, it is easy to heat them far above those temperatures. Moreover, low-temperature casting requires less-expensive equipment. Therefore, these metals are usually cast. This section discusses the properties and application of these two metals and their alloys.

ZINC

Pure zinc has a melting point of 787.3°F (419.6°C) and a hexagonal close-packed (HCP) crystal structure. It is a dense metal with a density of 7.14 g/cm^3, compared with 7.86 g/cm^3 for steel. Zinc has excellent atmosphere corrosion resistance and it adheres well on a substrate. The major application of zinc is as a protective coating for steels. A steel coated with zinc for corrosion resistance is called *galvanized steel*. Galvanized steel accounts for about 52 percent of the total zinc usage.

The bulk of the zinc which is not used for galvanizing is used in die cast operations. Die casting accounts for about 23 percent of the total usage of zinc. The low melting point of zinc alloys allows a wide variety of molds to be used for casting. The type of molds include plaster, metal, graphite, and even silicone rubber. The relatively low solidification temperature range for zinc alloys allows the part to be removed from the mold sooner than with aluminum or copper alloys.

Zinc is usually alloyed with aluminum for casting. Aluminum adds strength, reduces grain size, and minimizes the attack of the molten zinc alloy on the iron and steel casting equipment. Aluminum also increases the fluidity of molten zinc and improves its castability. The aluminum content ranges from 3.5 to 4.3 percent. At about 5 percent aluminum, a zinc-aluminum phase forms. This makes the alloy extremely brittle. The Zn-4%Al alloy, containing 4 percent aluminum, is typically used for casting. It has a high castability, is easy to finish, has good mechanical properties, and remains free of intergranular corrosion.

MAGNESIUM

Metallic magnesium has a density of 1.74 g/cm^3. It is the lightest of all commonly used structural metals. Magnesium alloys are used in similar applications as aluminum alloys. Applications for magnesium alloys include aircraft parts, missile parts, machinery, tools, material handling equipment, automobile parts, and computer parts. Magnesium can be found in large quantities in seawater. As methods for extracting this magnesium improve, magnesium and its alloys will be used more frequently.

Pure magnesium has a melting point of 1202°F (650°C). It has a hexagonal closed-packed (HCP) crystal structure. This crystal structure makes magnesium alloys, like zinc alloys, difficult to plastically deform at room temperature. Because of this, most magnesium alloys are used for die casting.

The tensile strength of pure magnesium at room temperature is about 27 ksi (186 MPa). Since this strength is insufficient for many structural uses, magnesium is usually alloyed with other metals. Aluminum is added to magnesium to increase strength, castability, and corrosion resistance. Examples of commercial magnesium-aluminum alloys are AM60 (Mg-6%Al) and AM100 (Mg-10%Al). AM60 is used for automobile wheels because of its lightweight and superior ductility. Zinc is also added to form finer precipitation particles of hardening phase. This increases strength. The Mg-Al-Zn casting alloys are important to industry because of their combination of light weight, strength, and relatively good corrosion resistance. For example, AZ91D (9.0 percent aluminum, 0.7 percent zinc) has a tensile strength of 34 ksi (234 MPa) and is used for parts in cars, lawn mowers, business machines, chain saws, hand tools, and sporting goods.

Besides being lightweight, magnesium alloys have excellent machinability. The machinability of magnesium alloys is 5 to 20 times better than that of AISI B1112 free machining steel. The atmospheric corrosion resistance of magnesium is not as good as aluminum, but it is better than steel. Chemical conversion coatings can be used to improve its corrosion resistance. Another concern with magnesium is its burn characteristics. Magnesium filings, powder, or chips burn extremely hot if ignited. In addition, the fire cannot be put out with water. Therefore, flood coolant must be applied in machining and grinding operations to prevent ignition.

PRECIOUS METALS

Precious metals are extremely important to modern civilization. High cost is what defines a precious metal. Jewelry, coins, bullion, and automobile catalytic converters are just a few of the everyday applications for precious metals. This section discusses the most widely used precious metals for industrial applications, including platinum, silver, and gold.

PLATINUM

Platinum (Pt) is primarily used in industry for corrosion resistance. Few materials can match the corrosion resistance of platinum in several high-oxidizing environments. In production of optical and other high-purity glass, platinum alloys are used for the lining of melting furnaces, pouring gates, and other parts where corrosion resistance is critical. Platinum alloys are the only materials that can be exposed in air to temperatures of about 2550°F (1400°C). Bushings and orifices of Pt-10%Rhodium are commonly used for producing glass fibers at this temperature.

Platinum is resistant to nitric acid at all temperatures and concentrations, including the extremely-high temperatures when boiling the acid under pressure. In these environments, platinum is greatly superior to stainless steels, which is commonly used with nitric acid at lower temperature. Platinum is also resistant to sulfuric acid at most temperatures and pressure. It is used to make vessels for storing concentrated sulfuric acid.

Platinum is also highly resistant to hydrogen fluoride and most fluoride compounds at high temperatures. Platinum-lined tubes are used in the production of organic fluorine compounds where temperatures reach 1300°F (705°C). Platinum containers are used in the production of large synthetic optical crystals from molten lithium fluoride and from molten sodium chloride.

Sulfurous gases at high temperature are destructive to most metals, including the common gold alloys and all silver alloys. Sulfur dioxide does not affect platinum. Hydrogen sulfide has only a slight effect on platinum.

Platinum has poor resistance to chlorine at elevated temperatures. However, at 212°F (100°C) and below it can be used. Above this temperature, particularly when the gas may not be dry, the specific environment must be considered before using platinum.

SILVER

The largest single use for pure silver is in photographic emulsions. The second largest use is in electrical and electronic applications. Silver or silver-based composites are used widely for electrical contacts for medium and high current/voltage applications. In these applications, high electrical conductivity, high resistance to oxidation, and arc erosion resistance are essential. Silver has the highest electrical and thermal conductivity among all metals, slightly higher than pure copper and much higher than gold. Unlike copper, silver does not easily oxidize at room temperature.

Many silver-based electrical contact alloys or composites have been developed including silver-cadmium oxide, silver-nickel, and silver-tungsten composites. For example, the silver-cadmium oxide composite is bonded onto a copper rivet for the electrical contacts used in automobile turn signal controls. These types of contacts are also used in limit and pressure switches, fractional horsepower motor start switches, relays, timers, and thermostats. See Figure 6-6. Silver particles of about 1 to 5 μm in diameter are commonly used in pastes for metallizing nonconducting materials.

Figure 6-6.
Silver cadmium oxide composite is used to make electrical contact tips for switches. The bases of the rivets are made of high conductivity copper. (Courtesy of CMW, Inc.)

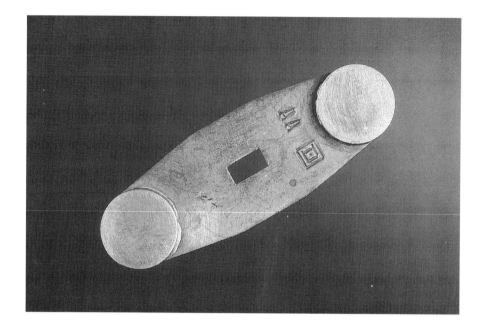

Silver has been used for many years in chemical and food processing equipment. These applications require high corrosion resistance and high product purity. In most of these applications, fine silver is used as a solid sheet, in tubular form, or as a lining over copper, nickel, or carbon steel plates. The silver linings can be bonded to the base metal by brazing or by solid-phase bonding. Silver is also used as a liner in heavy-duty journal bearings.

Silver is used extensively in applications where halogens and halogen acids are found. One important example is the handling of wet chlorine gas in water-purification installations. Control and metering equipment and tubing are constructed of fine silver or of a Au-7.5%Cu alloy in association with plastics.

Silver is also used in applications where aqueous solutions of hydrochloric acid are present. This is especially true when the acid is associated with organic hydrocarbon liquids, such as in pressure vessels and process equipment.

Alkaline cyanide in presence of air or other oxidizing agents rapidly dissolves silver. Nitric acid that contains traces of nitrous acid also attacks silver vigorously. Hot concentrated sulfuric acid, hot dilute sulfuric acid, and sulfur dioxide also attack silver or silver-based composites.

GOLD

In the past, gold was considered mainly a monetary metal. However, starting in the late 1950s, more gold was used by manufacturers and investors than was used for monetary purposes.

High-purity gold has a variety of applications. These include jewelry, dental uses, selective light filters, and thermal limit fuses. Gold is also used as target materials in x-ray equipment, a freezing-point standard, a high-melting solder to produce vacuum-tight pressure welds, linings for chemical equipment, and clad on phosphor bronze for contact springs in radio-frequency circuits.

Gold is also used extensively in electrical contacts. This is especially true for telecommunication equipment where oxidation resistance is critical to maintain proper signal-to-noise ratios.

Gold has many other applications in various industries. For example, it is used to provide a surface film resistant to corrosion and tarnishing in electrical applications. In electronics, gold is placed onto selected areas of silicon transistors and integrated circuit chips to produce electrical terminals. During this process, gold and silicon form a low melting compound that bonds the semiconductor to its base.

SUMMARY

This chapter has explored the broad spectrum of nonferrous metals and alloys for industrial applications. For example, pure copper is mostly used as an electrical and thermal conductor, whereas copper alloys, such as brass and bronze, are used for structural purposes. Aluminum is a light metal having good electrical and thermal conductivity and is often used in place of copper for applications with a weight limit. Aluminum alloys are mainly used as light structural materials finding many applications in the aerospace and construction fields. Titanium alloys are lighter than copper and iron, and offer an outstanding strength-to-weight ratio and corrosion resistance. They are particularly suited for components requiring high strength, but a low weight limit. Applications of zinc and magnesium are increasing due to their ease of fabrication. Precious metals are mainly used in industry for their superior corrosion and oxidation resistance. Silver and gold have good electrical conductivity and, therefore, are used in the electronic industry.

IMPORTANT TERMS

Age Hardening

Aluminum Bronze

Annealing

Anodizing

Beryllium Copper

Brass

Bronze

Cartridge Brass

Deoxidized Copper

Electrolytic Tough-Pitch (ETP) Copper

Galvanized Steel

Lead Brasses

Mains Metal

Nonferrous Metals and Alloys

Oxygen-Free Copper

Phosphorus Bronzes

Pitting

Red Brass

Silicon Bronze

Solution Treatment

Temper

Tin-Bronze

Yellow Brass

QUESTIONS FOR REVIEW AND DISCUSSION

1. What are nonferrous metals?
2. Which metal has the highest electrical conductivity?
3. Why is pure copper widely used as a conductor?
4. What is ETP copper?
5. What is oxygen-free copper?
6. What is another name for yellow brass?
7. What is the composition of red brass?
8. What is tin-bronze?
9. What is aluminum-bronze?
10. What is the alloy element in beryllium copper?
11. Why is copper the first choice for an electrical busbar (conductor)?
12. Why is brass stronger than pure copper?
13. Why is silicon bronze used for applications under seawater?
14. Why are special precautions required for casting beryllium copper?
15. What does T6 in the designation 6061-T6 stand for?
16. What is solution treatment?
17. What is age hardening?
18. Can aluminum be quench-hardened? Why or why not?
19. How is the modulus of elasticity of an aluminum alloy affected by the composition of alloying elements?
20. What are non-heat-treatable and heat-treatable aluminum alloys?
21. How can the surface of aluminum be modified to improve its corrosion resistance?
22. What safety precaution is required when machining magnesium alloys? Explain why.
23. What is the major characteristic of platinum?
24. Why is silver used as a material for electrical contacts?
25. What is the largest use of silver?

FURTHER READINGS

1. Smith, W.F. <u>Structures and Properties of Engineering Alloys (2nd ed.)</u>. New York: McGraw-Hill, Inc. (1993).
2. Smith, W.F. <u>Foundations of Materials Science and Engineering (2nd ed.)</u>. New York: McGraw-Hill, Inc. (1993).
3. Flinn, R.A. and Trojan P.K. <u>Engineering Materials and Their Applications (4th ed.)</u>. Boston: Houghton Mifflin Co. (1990).
4. <u>Metals Handbook, Vol. 3: Properties and selection—nonferrous alloys and pure metals</u>. Metals Park, OH: American Society of Metals (1978).

INTERNET RESOURCES

http://www.copper.org
  International Copper Association, Ltd.

http://www.aluminum.org
  The Aluminum Association, Inc.

http://www.titanium.com
  Titanium Industries, Inc.

http://www.titanium.org
  International Titanium Association

http://www.iza.com
  International Zinc Association

http://www.australianfoundries.com/magtech
  Magnesium Technologies Pty Ltd

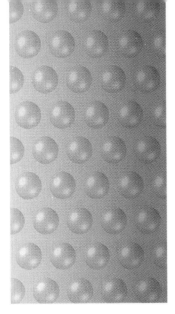

Chapter 7

Thermoplastics

KEY CONCEPTS

Upon completion of this chapter, you should understand:

> Classifications of plastics.
> Characteristics of thermoplastics.
> Definitions of polymers and polymerization.
> Thermoplastic processing.
> Types of thermoplastics.
> Viscoelasticity and glass transition of thermoplastics.
> Basic applications of thermoplastics.
> How and why plastics are recycled.

Since World War II, polymers have been the fastest growing segment of the chemical industry in the United States. No material has played a more important role in our daily lives as the synthetic polymers known as plastics. A modern automobile contains over 200 lbs (100 kg) of plastics. With the continued need to reduce vehicle weight to save fuel, plastics will continue to replace metals and alloys in the automobile industry. Plastics can be found everywhere in our homes. These applications include piping, residential flooring, siding, thermal and electrical insulation, decorative laminates, and so on. A trip through a supermarket quickly shows the importance of plastics in product packaging, such as bottles, films, and trays. On the basis of volume, plastic production in the United States has surpassed steel production. The current era is truly the "Plastic Age."

The major advantages that make plastics popular everywhere are their light weight, ease of fabrication, lower processing temperature, and lower cost. Moreover, most plastics do not rust like steels do.

For industrial applications, there are two basic types of plastics available. These are called thermoplastic and thermosets. *Thermoplastics* are polymers that soften when heated and regain their solid form when cooled. This process can be repeated multiple times. Because of this, thermoplastics can be ground up and reused for thermal processing. *Thermosets*, however, can pass through only one heat cycle. Heating causes thermoset to undergo a curing, or setting, reaction. Thus, thermoset cannot be re-melted or re-formed by a thermal process. This chapter introduces polymers and polymerization, polymeric structure, and the properties and applications of thermoplastics. Thermosets are discussed in the next chapter.

POLYMERS AND POLYMERIZATION

Thermoplastics are a type of synthetic polymer. To understand plastics, you must understand the concept of polymers. *Poly* means many and *mer* stands for monomer, or unit. A *polymer* molecule is made up of many repeating units called *monomers*. For example, the polymer, or plastic, used for making milk containers is known as polyethylene. The individual monomer in this polymer is ethylene (C_2H_4 or $CH_2 = CH_2$), as in Figure 7-1. If many ethylene molecules are bonded together, they form a chain. A polymer molecule can consist of hundreds of repeating units (monomers).

The process of linking monomers together is known as *polymerization*. Polymerization produces polymers or plastics having properties different from the corresponding monomers. For example, ethylene is a colorless and flammable gas at room temperature. However, polyethylene is a solid plastic used for making plastic grocery bags and milk jugs.

When repeating units of identical monomers are hooked together during polymerization, the resulting plastic is a *homopolymer.* For example, polyethylene is a homopolymer because it is made of ethylene monomers. When two or more types of monomers are included in the polymerization process, such as ethylene and propylene, the resulting polymer is a *copolymer.*

Figure 7-1 demonstrated the polymerization of ethylene to form polyethylene (PE). If one of the hydrogen atoms in ethylene is replaced by methyl (CH_3), it becomes propylene with 3 carbon atoms in monomer, as shown in Figure 7-2. Under certain conditions, the propylene can be polymerized to form *polypropylene* (PP). If one carbon in ethylene is substituted with chlorine (Cl), we have a new chemical known as *vinyl chloride*. The monomers of vinyl chloride can be polymerized to form polyvinyl chloride (PVC). By the same token, styrene is formed by substituting one hydrogen with a benzene ring in ethylene. Styrene can be polymerized into *polystyrene* (PS). These four thermoplastics—polyethylene, polypropylene, polyvinyl chloride, and polystyrene—belong to *polyolefin family* of polymers. Over two-thirds of industrial polymers used today are from the polyolefin family.

Different monomers in polymer molecules result in different materials properties. For example, polypropylene is formed through polymerization of propylene (C_3H_6). Polypropylene is stronger and tougher than polyethylene because of its larger monomer molecules.

Figure 7-1.
This is polyethylene, an example of a thermoplastic linear polymer molecule.

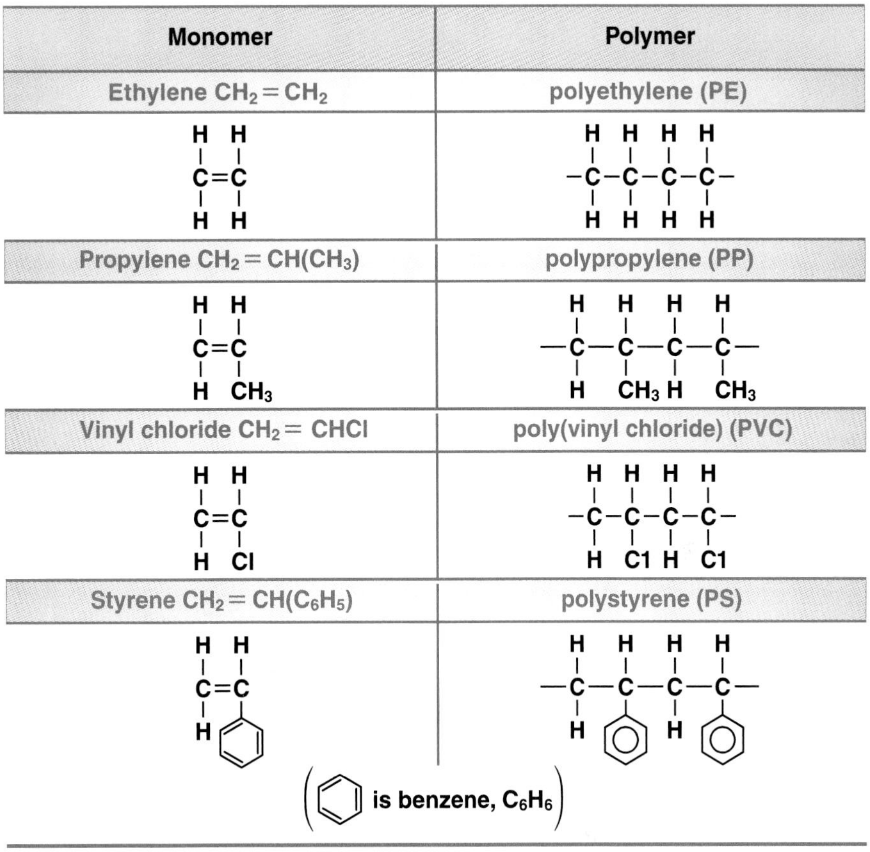

Figure 7-2.
Different polymers are formed by polymerization of their corresponding monomers.

Monomer	Polymer
Ethylene $CH_2 = CH_2$	polyethylene (PE)
H H | | C=C | | H H	H H H H | | | | −C−C−C−C− | | | | H H H H
Propylene $CH_2 = CH(CH_3)$	polypropylene (PP)
H H | | C=C | | H CH_3	H H H H | | | | −C−C−C−C− | | | | H CH_3 H CH_3
Vinyl chloride $CH_2 = CHCl$	poly(vinyl chloride) (PVC)
H H | | C=C | | H Cl	H H H H | | | | −C−C−C−C− | | | | H C1 H C1
Styrene $CH_2 = CH(C_6H_5)$	polystyrene (PS)
H H | | C=C | | H ⬡	H H H H | | | | −C−C−C−C− | | | | H ⬡ H ⬡

$$\left(\text{⬡ is benzene, } C_6H_6 \right)$$

POLYMERIC STRUCTURE

The property of any polymer is very dependent on the physical configuration of molecules. There are three basic configuration types for polymer molecules. These types are linear, branched, and network, Figure 7-3.

The simplest configuration type is a *linear* polymer, in which the length of a molecule is much larger than the breadth of the chain. Many thermoplastics have linear molecular distribution. Polymer chains may cross each other randomly in space, much like a bowl of spaghetti. Polymers with a linear molecular structure tend to be very flexible because linear polymer chains are not bonded together.

If a secondary chain grows at the same time the main chain extends during polymerization, a *branched* polymer molecule is formed. A branched polymer has branches, similar to a tree. Branching usually causes strengthening and stiffening of the polymer. Deformation of the polymer requires movement of chains that are more entwined than in linear polymers. Many elastomers or polymeric rubbers have a branched type of molecular structure. This is why they have such resilience and can withstand significant stretch without breaking.

Figure 7-3.
Polymer molecules exist in linear, branched, or network form. They can also exist as a hybrid of the three forms.

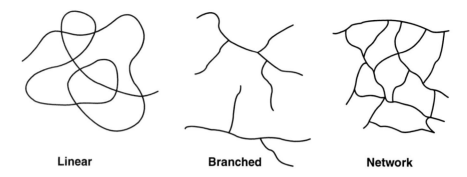

Linear   **Branched**   **Network**

If branches further grow to connect with other molecules, molecule chains cross-link and form a *network polymer.* Networked polymers usually cannot be remelted because the bonds between chains are too strong. The cross-link or network of molecules makes the polymer rigid, strong, and resistant to high temperatures. The more cross-linked the polymer is, the more rigid, less soluble, and higher temperature-resistant it is. Most thermoset plastics have a cross-linked structure and some can withstand temperatures as high as 400°F (200°C). On the other hand, linear polymers seldom can be used in temperatures over 250°F (120°C).

THERMOPLASTIC PROCESSING

Thermoplastic polymers as a raw material come in the form of pellets, granules, flakes, or powder. Most processes used to form thermoplastics involve first melting the thermoplastic. The melted plastic is then forced into a mold cavity or die. Finally, the plastic is cooled to harden it. Methods frequently used for processing thermoplastics include extrusion, injection molding, blow molding, and rotational molding.

EXTRUSION

Extrusion is a process for making a product by forcing a material through an orifice or die. It begins when the polymer is fed into the extruder through a feed opening, or hopper, in a pellet form. Then, the plastic is melted and forced through a forming die.

Figure 7-4 shows a single screw extruder. The screw is rotated by a drive motor through a gear reducer. The rotating screw forces the molten plastic toward the die. As the molten material is extruded from the die, a plastic piece with a profile of the die is formed. The extruded plastic is continuously shaped and cooled by equipment placed in the manufacturing line after the extruder. The equipment can include cooling rolls, water tanks, vacuum sizing fixtures, air cooling tables, pulling devices, cutting equipment, and coiling or winding equipment.

The main advantage of the extrusion process is its high efficiency of production. Products made by extrusion include plastic pipe, tubing, wire coating, plastic films and sheets, coatings for paper and foil, fibers, filaments,

Figure 7-4.
Thermoplastic particles are
fed through the hopper,
melted, and forced by the
screw through the extrusion
die to form the extruded
products.

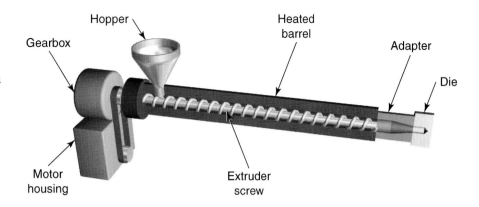

Figure 7-5.
Some examples of extruded
products. (Courtesy of DSM
Engineering Plastic Products)

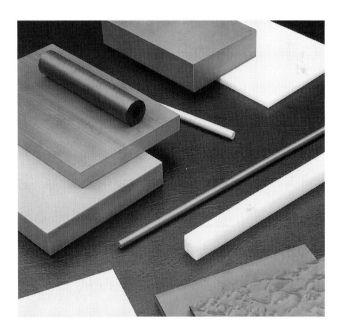

yarn tapes, and plastics plates. Extruded sheets and plates can be further
used for thermoforming to produce cups, buckets, and so on. Figure 7-5
shows some examples of extruded products.

INJECTION MOLDING

Injection molding is a major processing technique for making thermoplastic products. During *injection molding,* molten plastic is forced into a
mold cavity to form a product. Figure 7-6 shows a typical reciprocating
screw injection molding machine. The screw is usually turned by a
hydraulic motor. Granular thermoplastic resin is fed from the hopper into
one end of the heated barrel. The thermoplastic raw material is heated and
melted. The melted plastic is then forced by the screw into the injection
chamber at the front end of the barrel. The screw rotation is stopped before
the melted plastic is injected into the mold cavity.

Figure 7-6.
During the injection molding process, solid thermoplastic particles are fed through a hopper into a heated extruder barrel. The molten material is transported into a mold by the rotating screw, followed by the plunger action of the nonrotating screw.

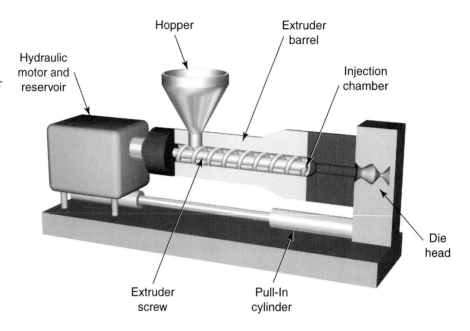

Injection cylinders are used to inject the melted plastic into the mold. The mold is held closed by a clamping mechanism. The plastic cools and hardens inside the mold until it is fully set. Finally, the mold opens and the molded part is ejected. Injection molded products can have various shapes and sizes, depending on the design of the mold cavity.

BLOW MOLDING

Blow molding utilizes compressed air to inflate molten plastic to conform to the mold cavity. It is a major way of making hollow containers such as bottles. Glass blowing, common in ancient times, is an example of blow molding. The blow molding of thermoplastic materials began during World War II. Polystyrene was the first material used with the newly developed blow molding machines. Low-density polyethylene was used in the first large-volume commercial application—a squeeze bottle for deodorant. Today, polyethylene resins are still primarily blow molded. About 85% of all plastic bottles on the market today are made from blow molded high-density polyethylene.

Generally, the blow molding process consists of three stages. The first stage is to melt thermoplastic resin. The second stage is to form a cylindrical tube known as a *parison.* The third stage is to blow air into the parison to force the plastic to fit the mold. When the piece is solidified, the mold is opened and the finished part is ejected. In *extrusion blow molding,* as shown in Figure 7-7, a parison is extruded downward. The mold closes around the parison, pinching off the bottom of the parison and forming threads for the neck of the bottle. Then, the parison is inflated by an air blast. After the plastic solidifies, the part is ejected from the mold and the next parison is extruded into the mold.

Figure 7-7.
Extrusion blow molding machine and steps of blow molding: 1) melting thermoplastic resin, 2) formation of a parison by extrusion, and 3) blowing the parison into the desired shape.

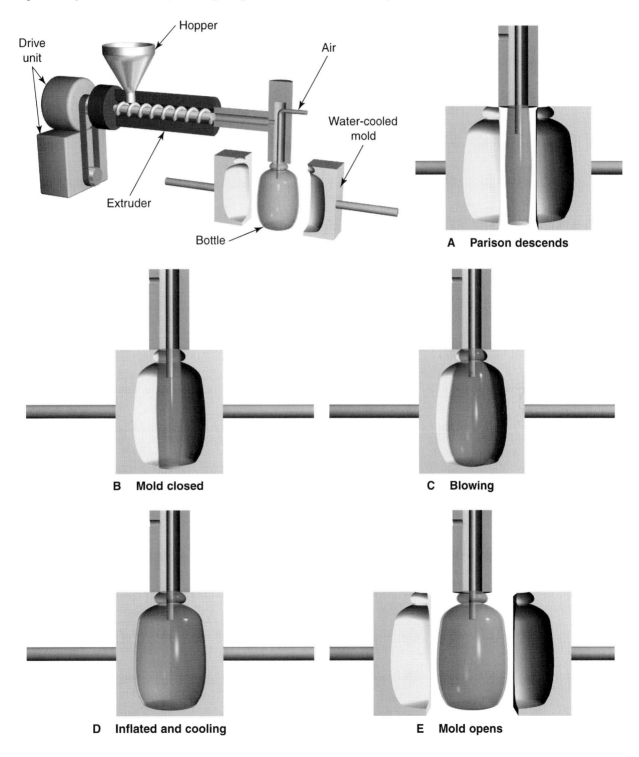

Most blow molding applications are containers. Hollow-skinned plastic items such as bleach bottles, liquid soap dispensers, and gas tanks for vehicles have been produced. Almost any container can be produced by blow molding. Some containers may have handles. Other containers may have molded-in spouts. Very large containers can be produced. Large, blow molded containers are often used in the agriculture industry. Many farms use these containers to hold chemical fertilizer for plants or water for livestock. Some of these containers can hold 1000 gallons or more.

ROTATIONAL MOLDING

Rotational molding, or *centrifugal casting,* utilizes a rotating mold cavity to form a hollow product. It has become an important alternative process to injection molding, and blow molding. Since the early 1960s, high- and low-density polyethylene began to be extensively used in rotational molding. Today, polyethylene is the most frequently used plastic for rotational molding. Other engineered thermoplastics are used in various applications as well.

In rotational molding, a charge of fine thermoplastic powder is placed in a heated mold. The mold is then rotated about two perpendicular axes. This rotation distributes the powder over the inner mold surface while at the same time, the particles are melted. The mold is then cooled by compressed air or a water spray, opened, and the part is ejected. Because of the characteristics of the process, rotational molding is capable of producing extremely irregular hollow objects. The molds are inexpensive to produce. Often, a mold can simply be formed from sheet steel because high pressures are not involved in the rotational molding process.

CLASSIFYING THERMOPLASTICS

The types of thermoplastics are greatly varied. In addition, new developments in thermoplastics are emerging quickly. Therefore, it is impractical to cover all types of thermoplastics. This section briefly introduces important and commonly used thermoplastics. The thermoplastics covered in this section include the polyolefin family, styrenic plastics, vinyl plastics, and acrylics. These polymers are widely used commercially, with millions of tons produced annually. The other types of thermoplastics covered in this section are called engineering resins or engineering thermoplastics. These are typically used in place of metals rather than glass, paper, and wood. Engineering thermoplastics are produced in much smaller quantities than the other thermoplastics and are priced higher.

POLYOLEFINS

Polyolefins are low cost thermoplastics. The polyolefin family of thermoplastics includes polyethylene (PE), polypropylene (PP), polybutylene, polymethylpentene (PMP), and ethylene-vinyl alcohol (EVOH). They are the most commonly used industrial polymers.

Polyethylene (PE)

Polyethylene is characterized by its high toughness, near-zero moisture absorption, excellent chemical resistance, superior electrical insulating properties, low coefficient of friction, and ease of processing. Polyethylene is usually classified by its density, as follows.

- Ultra- or very-low density: 0.880 to 0.915 g/cm^3
- Low density: 0.910 to 0.925 g/cm^3
- Medium density: 0.926 to 0.940 g/cm^3
- High density: 0.941 to 0.965 g/cm^3

The primary differences among these types of polyethylene are rigidity, heat resistance, chemical resistance, and ability to sustain loads. In general, as the density increases, the hardness, heat resistance, stiffness, and resistance to moisture absorption increase.

There are two special types of high density polyethylene (HDPE). The first is high molecular weight, high density polyethylene (HMW-HDPE). The second is ultra-high molecular weight, high density polyethylene. *Molecular weight* is the molecular weight of the monomer multiplied by number of monomers in the polymer molecule. In other words, it is proportional to the molecular chain length. The higher the molecular weight, the longer the molecular chain for linear polymers. Generally, increasing molecular weight increases the rigidity, tensile strength, and wear resistance of the polymer.

Polypropylene (PP)

Polypropylene is currently the most versatile of the family. Its use continues to increase. First produced in the 1950s, early polypropylene suffered from low yields in polymerization and poor control of molecular weight and, therefore, its properties. The development of high-activity catalysts in the 1970s improved the yields dramatically and made the thermoplastic cost effective.

Polypropylenes have better resistance to heat and resist more chemicals than many other similarly-priced thermoplastics. Moreover, polypropylenes have negligible water absorption, excellent electrical properties, and are easy to process.

Polybutylene

Polybutylene resins are linear and flexible polyolefins offering a unique combination of properties. Polybutylene is made from polyisobutylene, a distillation product of crude oil. Its monomer molecule has two methyl groups (CH_3) substituted for hydrogen atoms in the basic ethylene molecule, as in Figure 7-8.

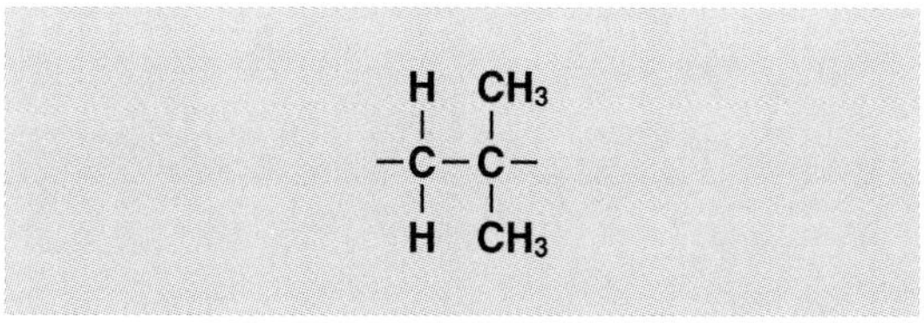

Figure 7-8.
A monomer of polybutylene.

Polymers tend to stretch without an increase in applied load. This is known as *creep.* Some of the important characteristics of polybutylene are excellent resistance to creep at elevated and room temperatures, good toughness, and exceptional resistance to environmental stress cracking. *Environmental stress cracking* occurs when certain materials are exposed to stress and environmental media. Environmental stress cracking usually causes fracture catastrophically without any prior warning.

Polymethylpentene (PMP)

This thermoplastic essentially contains a short carbon atom chain as a substitute on an ethylene molecule. Polymethylpentene is characterized by transparency, low density of 0.83 g/cm^3, and a high melting point of 464°F (240°C). Its drawbacks are brittleness and poor ultraviolet resistance.

Ethylene-Vinyl Alcohol (EVOH)

Ethylene-vinyl alcohol (EVOH) provides an excellent barrier to the permeation of oxygen and other gases. It also can prevent permeation of odors and flavors, which is useful for food packaging. However, EVOH has a strong tendency to absorb moisture. Therefore, it is generally used inside a package made of polyethylene.

STYRENIC PLASTICS

Styrenic plastics are a family of polymers related to styrene. In general, styrenic plastics are characterized by ease of processing, hardness, and excellent dielectric properties. They are relatively limited in heat resistance and are attacked by certain hydrocarbon solvents.

Polystyrene (PS)

The molecular configuration of polystyrene is shown in Figure 7-2. General-purpose crystal polystyrene is the low-cost member of the family. Properties of the amorphous polystyrene include high hardness, rigidity, optical clarity, dimensional stability, and easy to process. Most crystal polystyrene is injection-molded.

Impact polystyrene is formed with butadiene elastomers. This extends the use of polystyrene into areas where high impact strength and good elongation are required. High-impact polystyrene is the best styrenic plastic for load bearing applications.

There are other types of polystyrene as well. These types offer various improvements in chemical, thermal, and optical performance. However, desirable mechanical properties usually have to be sacrificed to achieve these other improvements.

Styrene-Acrylonitrile (SAN)

Styrene-acrylonitrile (SAN) copolymers are transparent, amorphous materials with higher heat and chemical resistance than crystal polystyrene. SAN can be injection-molded, but needs to be dried before processing.

When modified with an olefin, SAN becomes a tough, weather-resistant polymer that can be extruded into sheets or profile, as well as injection-molded.

ABS Plastics

ABS plastics are combinations of acrylonitrile, butadiene, and styrene in varying proportions. ABS plastics were originally an outgrowth of polystyrene modification to improve its toughness by copolymerization with butadiene-acrylonitrile rubber. A wide range of properties can be achieved by different proportions of the three monomers. Commercially, ABS plastics are available in a variety of grades, including medium- and high-impact, heat-resistant, fire-retardant, and low- and high-gloss types.

ABS plastics provide a balanced combination of mechanical toughness, a wide range of service temperatures, chemical resistance, electrical insulating properties, and ease of processing. In general, ABS plastics have very good resistance to a wide range of chemicals. They exhibit low water absorption. ABS plastics generally stand up well in weak acids and, generally, in strong acids. However, they deteriorate in solvents such as ester, ketone, and some chlorinated hydrocarbons.

VINYL PLASTICS

Vinyl plastics include a family of polymers related to vinyl chloride. See Figure 7-2. This family normally decomposes at high temperatures. Thus, stabilization of the polymer is an important issue.

Polyvinyl Chloride (PVC)

Polyvinyl chloride (PVC) is one of the most widely used plastics. Its acceptance comes from its versatility. In addition, it can be used in rigid compounds or blended with a plasticizer to produce flexible grades.

Rigid polyvinyl chloride normally contains PVC resin, a heat stabilizer, and an impact modifier, such as ABS or chlorinated polyethylene. Flexible PVC compounds often contain plasticizer to soften the plastic.

PVC degrades relatively easily and the degradation products are corrosive with chlorine. Therefore, processing of PVC requires special care. Machinery surfaces in contact with the polymer should be corrosion resistant. Processing temperature should be limited to as low as possible.

PVC resin will ignite, although it is self-extinguishing. However, hydrochloric acid and other toxic and corrosive chemicals are produced when PVC resin is burned. Therefore, only a properly designed incinerator should be used to dispose of PVC.

Polyvinylidene Chloride (PVdC)

Polyvinylidene chloride (PVdC) is a copolymer of vinylidene chloride and vinyl chloride or other monomers. These materials exhibit exceptional resistance to permeation of oxygen, carbon dioxide, water, and many organic solvents. It is an ideal material for food wrap and medical packaging.

ACRYLICS

Acrylic is the generic name of polymethyl methacrylate (PMMA). This rather complicated name actually relates to a plastic everyone should know. Its trade name is Plexiglas®. PMMA is a hard, rigid, and transparent plastic used to make low cost lenses, windows, and safety glasses.

Acrylics, or PMMA, are polymers obtained by reacting methylacrylic acid with an alcohol. The molecular configuration for PMMA monomer is shown in Figure 7-9. The monomer has methyl group (CH_3) replacing one hydrogen atom in an ethylene molecule and a $COOCH_3$ constituent replacing another hydrogen atom.

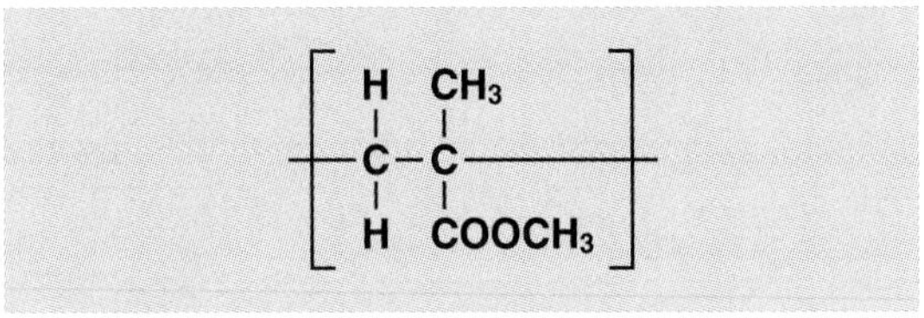

Figure 7-9.
A molecule of acrylics or polymethyl methacrylate (PMMA).

ENGINEERING THERMOPLASTICS

While there are fewer engineering thermoplastics than the other types, the number is increasing very fast and their applications are very promising. This section briefly discusses four of the typical engineering thermoplastics.

Acetals

Acetals are polymers of formaldehyde, known as polyoxymethylenes (POM). They are among the strongest and stiffest of the thermoplastics. Acetals are also characterized by excellent fatigue resistance and dimensional stability. Other outstanding properties include low friction coefficients, exceptional solvent resistance, and high heat resistance for extended use up to 220°F (105°C).

Fluoroplastics

Outstanding properties of fluorocarbon polymers or fluoroplastics include being inert to most chemicals and a resistance to high temperatures. Fluoroplastics have a waxy feel, extremely low coefficient of friction, and excellent dielectric properties. However, their mechanical properties are usually low.

Many fluoroplastics are available. The most common is polytetrafluoroethylene (PTFE) commercially known as Teflon®. The monomer has one fluorine atom replacing each hydrogen atom on an ethylene molecule, as in Figure 7-10.

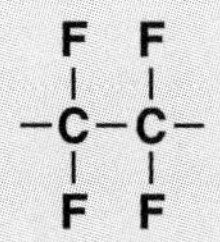

Figure 7-10.
This is polytetrafluoroethylene
(PTFE).

Polyamides (Nylons)

There are a number of common polyamides, or nylons. The nylon family members are identified by the number of carbon atoms in the monomers. A typical monomer molecule of nylon 6 is shown in Figure 7-11. This nylon 6 molecule has six carbon atoms in the polyamide molecule.

Nylon has excellent mechanical properties. However, it is affected by the moisture content of the resin during processing. Moisture control is a critical consideration for quality assurance of nylon products. Nylon has very good resistance to fatigue and creep. Its impact strength is high. It also has a low friction coefficient and high wear or abrasion resistance.

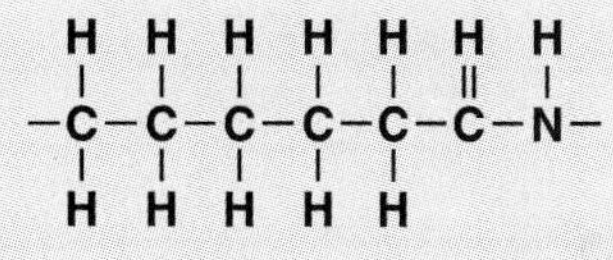

Figure 7-11.
A typical monomer molecule
of nylon 6.

Polycarbonate

Polycarbonate is an engineering polymer of increasing importance in many areas of industry, including medical devices and safety equipment. Figure 7-12 illustrates typical molecular structure of polycarbonate. It has a large, complex molecule and the polymer is strong and rigid. Polycarbonate is a clear (transparent) polymer. Its mechanical strength is close to crystalline nylon and acetal. Moreover, polycarbonate has exceptionally high impact resistance, good electrical properties, good thermal stability, and good creep resistance. Its impact resistance makes it one of the primary materials used in motorcycle helmets. Polycarbonate can be processed with all conventional thermoplastic processing methods.

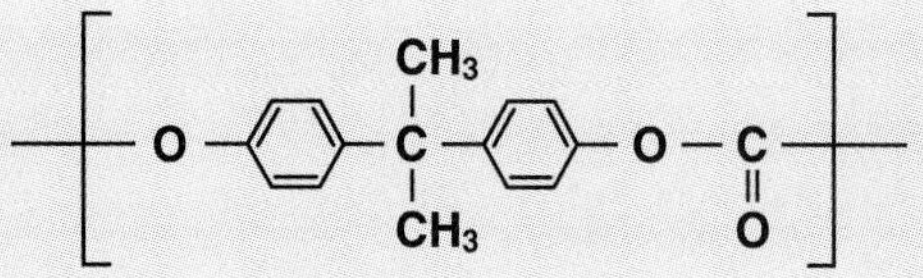

Figure 7-12.
A monomer of polycarbonate.

PROPERTIES

In this section, the special mechanical properties of thermoplastics are discussed. First, the tensile behavior of polymers in general is discussed and compared to metallic materials. Then, two special problems with thermoplastics—viscoelasticity and glass transition—are discussed.

STRESS-STRAIN BEHAVIOR OF THERMOPLASTICS IN TENSION

A tension test can be performed on thermoplastics in the same fashion as for metallic materials. Refer to Chapter 3 for a discussion of mechanical properties. These properties apply to both metallic and thermoplastic materials. Thermoplastics differ considerably for one another in modulus of elasticity, yield stress, and elongation prior to fracture. Figure 7-13 shows tensile stress-strain curves for polyvinyl chloride (PVC) and polystyrene (PS). The modulus of elasticity is measured by the slope of the linear portion of the stress-strain curve. These typical thermoplastics have varying degrees of ductility. In the two examples, polystyrene has higher modulus of elasticity than polyvinyl chloride. This means that PS is more rigid than PVC. Also, PS has higher yield strength and ultimate tensile strength than PVC. However, PS has much less plastic deformation or elongation prior to fracture than PVC, as measured by strain before fracture. That is, the ductility of PVC is much greater than that of PS.

Stress-strain behavior for plastic is similar to that of metals when the test is performed at room temperature and in a short period of time. However, polymer properties change significantly over time.

Figure 7-13.
This figure shows two tensile stress-strain curves for two thermoplastics—polyvinyl chloride (PVC) and polystyrene (PS). The specimen deformations at various points of stress-strain curve similar to metallic materials.

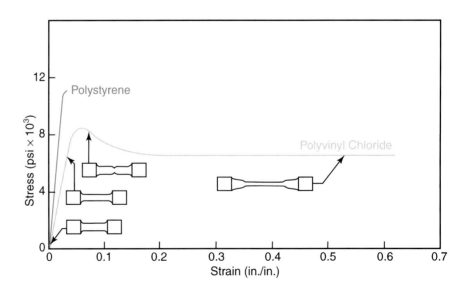

VISCOELASTICITY

One of the most striking characteristics of polymers is that their properties are highly dependent on time. In natural world, there are elastic solids, such as steels, and viscous fluids, such as oil. These are the extremes of a broad spectrum of material response. Polymers fall somewhere between. The mechanical behavior of polymers has certain material responses similar to elastic solids and to viscous fluids. This is known as *viscoelastic behavior.*

To understand the viscoelastic behavior of polymers, first look at the strain-time relationships for perfectly elastic solids and viscous fluids. Then, compare those relationships with polymers. Figure 7-14 illustrates the typical responses of the three materials.

A constant amount of load is applied at time t_1 and removed at time t_2. As shown in Figure 7-14A, elastic deformation is instantaneous and does not change with time for an elastic solid. After the stress or load is removed from the elastic solid, all the deformation is reduced to zero.

For a viscous liquid, there is a linear increase in deformation with time while a constant stress is applied at time t_1, as shown in Figure 7-14B. Unlike elastic solids, the deformation is completely irrecoverable. The strain or deformation is permanent at the level, even if the stress or load is released at time t_2.

A typical polymer possesses a combination of elastic and viscous properties. As in Figure 7-14C, a polymer is first deformed elastically if a constant load or stress is applied (s/G_1 at t_1). With a constant load, the polymer continues to deform as time increases. After the load is removed at time t_2, the strain decreases to certain level, but not completely to zero. In other words, a polymer has some delayed deformation occurring after the stress is applied and kept constant. This delayed material response to stress is known as *viscoelasticity.*

Figure 7-14.
Strain-time relationships of constant stress for: A—ideal elastic solid. B—ideal viscous solid. C—typical polymer showing viscoelasticity.

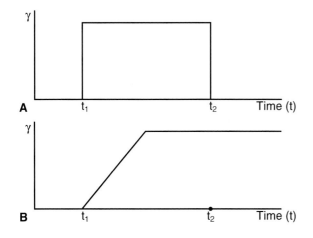

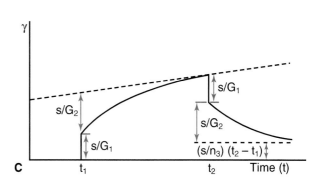

Viscoelasticity of polymers creates two practical application problems for thermoplastics. First, if a constant stress or load is applied to a polymer product, the deformation continues to increase with time. This phenomenon is known as *creep*. Creep causes excessive deformation on the part. Therefore, the part loses its dimensional stability. Secondly, if a polymer is deformed and held at some level of strain, the applied stress required to maintain the same amount of strain decreases over time. As time increases, the plastic part loses its loading capability. This is known as ***stress relaxation.*** To compensate for these two problems, a much larger safety factor is normally used in the design of plastic products than for metallic products.

GLASS TRANSITION

Amorphous polymers can exhibit two distinctly different types of mechanical behavior. Some polymers, like acrylic and polystyrene, are hard, rigid plastics at room temperature. Those polymers are brittle like glass. Other polymers, like polybutadiene and polyethyl acrylate, are soft, flexible, and rubbery plastics. However, a rigid plastic becomes rubbery if it is heated above certain temperature. For example, if polymethyl methacrylate and polystyrene are heated to about 255°F (125°C), they exhibit typical rubbery properties. On the other hand, rubbery plastic becomes rigid and brittle if it is cooled below certain temperature. When a rubber ball is cooled in liquid nitrogen, it becomes rigid and shatters when you try to bounce it. The change from rubbery state to glassy state is known as ***glass transition.***

For any polymer there is a temperature below which it is in a glassy state and above which it is rubbery. This temperature is known as ***glass transition temperature (T_g).*** The glass transition temperature is a property of the plastic. Whether the polymer has a rigid or a rubbery property depends on whether its application temperature is above or below the polymer's glass transition temperature.

The glass transition temperature of a polymer is very important to consider when selecting plastics for an application. Figure 7-15 lists some glass transition temperatures for various thermoplastics. Since the

Figure 7-15.
This table shows the glass transition temperature for various thermoplastics.

GLASS TRANSITION TEMPERATURES FOR THERMOPLASTICS		
Polymer	**Glass transition temperature (Tg)**	
	°F	°C
Polyethylene (PE)	−193	−125
Polypropylene (PP)	14	−10
Polyvinyl fluoride (PVF)	−4	−20
Polyvinyl chloride (PVC)	199	87
Polystyrene (PS)	212	100

glass transition temperature for polyethylene (PE) is –193°F (–125°C), polyethylene is soft and flexible at room temperature. On the other hand, polystyrene (PS) is rigid at room temperature because its glass transition temperature is 199°F (87°C). Boiling water placed in a polystyrene cup will "melt" the cup.

APPLICATIONS

Plastics are finding more and more applications in all corners of our lives. On the basis of volume, plastics production in the US has surpassed steel production. The increase of plastic applications is expected to continue. This is especially true for thermoplastics because of their low processing cost. This section focuses on typical applications of the thermoplastics discussed earlier.

POLYOLEFINS

Polyolefins are the most used of all plastics. They are low in resin cost and generally produced in large quantities. The applications of polyolefins have changed the way we live.

Polyethylene (PE)

A few traditional markets for low-density polyethylene (LDPE) are in packaging films, extrusion coating of paper, wire and cable coating, injection molded parts, piping, and tubing. Linear low-density polyethylene (LLDPE) polymers, with little long-chain branching, have much greater elongation than LDPE. Their high tear, tensile, and impact strength, along with improved resistance to environmental stress cracking, allow stronger products to be produced with less material. This has been particularly important in film markets where films are being made thinner.

High-density polyethylene (HDPE) is highly crystalline and tough polymer that can be formed by most processing methods. A significant amount of HDPE is blow-molded into containers for household and industrial chemicals. It can also be injection-molded into items such as crates, housewares, and pails. It can be extruded into film for packaging. It can also be rotationally molded into trash containers, toys, and sporting goods.

High molecular weight high-density polyethylene (HMW-HDPE) is made into "blow film" for packaging, extruded into pressure pipe, and blow-molded into large shipping containers. Extruded sheets are used to form truck bed liners and pond liners.

High strength, excellent chemical resistance, and low friction make ultra-high molecular weight high-density polyethylene (UHMW-HDPE) ideal for gears, slides, rollers, and other industrial parts. It is also used to make bearings for artificial replacement joints, as in Figure 7-16.

Figure 7-16.
Ultra-high molecular weight
high-density polyethylene
(UHMW-HDPE) is used for
replacement joint bearings.
(BIOMET)

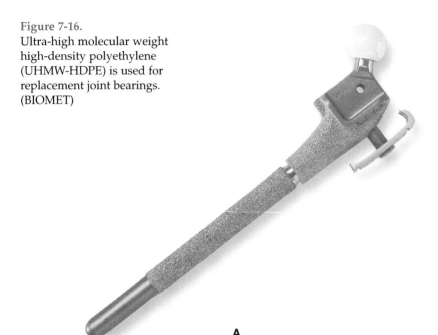

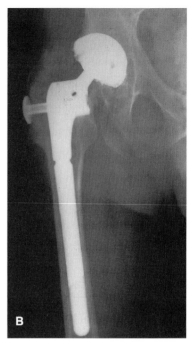

A

B

Polypropylene (PP)

The greatest commercial use for polypropylene (PP) is in fibers and filaments. PP fibers are woven into fabrics and carpets. They are also used to produce nonwoven fabrics for disposable diapers. Moreover, PP is made into unoriented and oriented films for packaging, which have largely replaced cellophane and glassine. Homopolymer PP is injection-molded into caps and closures, appliance components, and auto parts.

Copolymer polypropylene (PP) with up to 7 percent ethylene has higher impact strength and better clarity than the homopolymer PP. PP impact copolymers are tough even at low temperatures, and yet retain a high percentage of stiffness. Injection molding applications include automobile battery cases, interior and exterior trim parts and housewares. Co-extruded with barrier polymers such as ethylene-vinyl alcohol (EVOH), impact PP is made into a multilayer sheet thermoformed into food packages that can withstand freezer storage and microwave cooking.

Polybutylene

The most important commercial application for polybutylene is in plumbing pipe for residential and commercial use. Large-diameter polybutylene pipe is also used for transporting abrasive fluids. In sheet form, it is used for vessel and chute linings.

Polybutylene is used to make hot melt adhesives and sealants and as a carrier resin for additive and color concentrates. The resin is also utilized as a modifier for other polyolefins. For example, it improves processability, impact, and weld line strength for polypropylene. Polybutylene also improves processability and stress crack resistance for polyethylene.

Polymethylpentene (PMP)

Polymethylpentene (PMP) is a thermoplastic with good optical clarity and can be used for high temperatures up to 300°F (150°C). It is used to make injection-molded and blow-molded laboratory ware and medical products, food processing equipment, and microwavable packaging. The combination of transparency, high-temperature resistance, and good electrical properties make this plastic suitable for lighting and appliance components.

Ethylene-Vinyl Alcohol (EVOH)

As excellent barriers to oxygen and other gases, odors, and flavors, ethylene-vinyl alcohol (EVOH) is incorporated in multilayer composite structures by coating, laminating, co-extrusion, or co-injection molding. The major uses of EVOH barrier packaging are food products although there are applications involving solvents and other chemicals. The most common combination is with polypropylene. Blow-molded bottles for ketchup, barbecue sauces, and salad dressings, and thermoformed containers for jellies, sauces, and prepared foods are all applications of EVOH.

STYRENIC PLASTICS

Styrenic plastics are commonly used as consumer products and packaging. This is due to their ease of processing and low cost. Moreover, styrenic plastics possess useful properties such as optical clarity and expandability.

Polystyrene (PS)

Most crystalline polystyrene is injection-molded. Applications include packaging for food, cosmetics, and pharmaceuticals; audio cassette cases; dust covers; disposable drinking cups; and cutlery. Typical applications of high-impact polystyrene modified with polybutadiene elastomers include food packaging such as cups, lids, and takeout containers; housewares; office products; and video cassettes. Combustion-resistant grades are used in radio and TV housings.

Foamed crystal polystyrene has cushioning and insulation properties at very low densities. Extruded foam sheets are used as a protective wrapping material and is thermoformed into fast food packages, egg cartons, and meat and produce supermarket trays. Expandable polystyrene (EPS) is supplied as small beads containing a blowing agent. The beads are pre-expanded with steam to establish the density of the finished product and then steam-chest-molded into lightweight products such as disposable coffee cups, cooler chests, and protective package inserts.

Styrene-Acrylonitrile (SAN)

Styrene-acrylonitrile (SAN) is injection-molded into such products as dishwasher-safe housewares, refrigerator shelves, medical devices, connectors for PVC tubing, and lenses. Olefin-modified SAN is used as an extruded capstock (protective shield) for ABS and other less expensive substrates to provide weatherability for spas, boats, swimming pools, and recreational vehicles.

ABS Plastics

There is an extremely wide range of applications for ABS plastics. Falling between the commodity plastics and the engineering resins, ABS meets property requirements for many parts at a reasonable price. In the automobile industry, ABS is used to injection-mold interior panels and trim, grilles, wheel-covers, and mirror housings. ABS is often used to blow-mold bumper fairings for large trucks.

Flame-retardant grades are used to mold housings and keyboards for computers. General purpose ABS is used for computer floppy disk housings. It is also used to mold parts for telephones, calculators, business machines, and appliances. ABS sheet is thermoformed into refrigerator door liners and food storage compartments. Other ABS applications include housewares, luggage, toys, and sporting goods.

VINYL PLASTICS

Vinyl plastics are very versatile materials. Their properties can be modified to suit various applications. They can be produced in a rigid form. They can also be produced as a flexible grade by blending plasticizers into the plastic.

Polyvinyl Chloride (PVC)

Almost 75% of rigid polyvinyl chloride (PVC) produced is used for building and construction applications. Most PVC is processed via extrusion into products such as pipe, siding, and window profiles. Packaging is another market for PVC. Rigid grades are blown into bottles and made into sheets for thermoforming boxes and blister packs. Flexible compounds are made into food wraps. Other markets for PVC include wire and cable coating, flooring, garden hoses, and toys.

Polyvinylidene Chloride (PVdC)

As an excellent barrier to oxygen, carbon dioxide, water, and many organic solvents, polyvinylidene chloride (PVdC) is made into monolayer food films (plastic wrap) and medical packaging. It is also used in co-extruded film and sheet structures as a barrier layer. PVdC coatings are applied to containers to prevent gas transmission. However, this layer has to be removed before the packages are recycled.

ACRYLICS

Acrylics have outstanding resistance to long-term sunlight exposure and weathering. Typical applications include outdoor glazing, aircraft canopies, skylights, automobile tail lights, dials, buttons, lighting applications, knobs, and machine covers. Some transparent acrylic multipolymers are also used in the drug and food packaging industry.

ENGINEERING THERMOPLASTICS

Engineering thermoplastics are much more expensive than commodity plastics. However, they have superior properties over other plastics. Engineering thermoplastics are generally used in the same applications as for metals.

Acetals

Acetals are available as compounds for injection molding, blow molding, and extrusion. Filled, toughened, lubricated, and ultra-violet-stabilized versions are available. Grades reinforced with glass fibers, fluoropolymer, and aramid fibers also are on the market. Glass fibers are used to achieve higher stiffness, lower creep, improved dimensional stability. Aramid fibers are used to improve frictional and wear properties.

Many applications involve using acetals in place of metals. This is done where higher strength of metals is not required and costly finishing and assembly operations can be eliminated by using the plastic. Typical parts include gears, rollers, bearings, conveyor chains, automobile door handles and window mechanisms, plumbing components, and pump parts.

Fluoroplastics

Polytetrafluoroethylene (PTFE) was the first, and still most common, fluoroplastic. PTFE, or Teflon®, is one of the most chemically inert thermoplastics known. Therefore, it is used for seals, tubing, and small vessels made for very aggressive chemicals. PTFE also has extremely low coefficient of friction. Major limitations of PTFE are it is expensive, not moldable by conventional techniques, has low strength, and has low creep resistance.

The fluoroplastic is usually supplied in powder form for compression molding or in water-based dispersions for coating and impregnating. It should not be used as structural components to carry loads.

Polyamides (Nylons)

Polyamides (nylons) are mainly used for textiles. Historically, nylon's first market was the silk hosiery market. Hence, ladies hosieries are now often generically known as "nylons." After overwhelming acceptance of nylon hosiery, the fabric and home furnishings markets were penetrated. This was mainly because of nylon's excellent knittability and durability. Industrial applications, especially in carpeting, followed. Besides the fabric applications, nylon has also found many uses where toughness is a prerequisite. Some examples include oil filler caps for motor vehicles, teeth in plastic zippers, castors (wheels) for light furniture, and automobile radiator tanks.

General-purpose nylon is available for extrusion, injection molding, blow molding, rotational molding, and casting. It is widely used for gears, cams, and slides on machinery. Nylon is also available as sheets and films.

For specific engineering applications, a number of specialty nylons have been developed. Molybdenum disulfide is used as filler in the nylon matrix to improve wear and abrasion resistance, frictional characteristics, flex strength, stiffness, and heat resistance. Glass fibers are added to nylons to improve tensile strength, heat distortion, and impact strength.

Nylons are also increasingly used as alloying elements for other polymers, notably with ABS plastics and polyethylene ether. Polyethylene ether is a polymer similar to polystyrene. Nylon's aroma barrier properties have led to applications in multilayer packaging films. In many electrical applications, nylon's mechanical strength and resistance to oils and greases are also important properties.

Polycarbonate

Polycarbonate is superior in many aspects to metal and glass. Therefore, it is increasingly used in place of those materials. Examples of applications include headlamp and vehicle signal lenses, tractor window frames, automobile instrument panels, sun room "glass," computer cases, compact discs, drinking pitchers, and refrigerator storage bins. See Figure 7-17. Other application examples are electronic and telecommunication instruments, medical syringes, optical lenses, packaging, and photographic equipment. Polycarbonate is also the material of choice for motorcycle and other safety helmets.

PLASTICS RECYCLING

A major concern with the use of plastics for disposable items is the impact on the environment. A study for US Environmental Protection Agency (EPA) revealed that an average of 4 pounds of trash is produced daily by every man, woman, and child in the US. Because of the increase in amount of solid waste and the decrease in the number and size of landfills, major cities are running out of landfill space. It is estimated that plastics occupy about 20% of landfill volume.

Plastic packaging is typically discarded within one year of manufacture. In some cases, this time period is even shorter. Besides packaging, plastics are used for all sorts of other applications. Therefore, it is increasingly important to recycle and reuse plastics as a way to decrease the amount of solid waste going to landfills. To help achieve this, many cities

Figure 7-17.
Polycarbonate is used to make these refrigerator storage bins. (Courtesy of Miles)

have recycling programs to collect paper, metal, and plastic. In addition, researchers are trying to understand the property change due to repeated processing and recycling of plastics, and to find means to improve properties of recycled plastics.

Thermoplastics can be remelted and reprocessed. Thermoset plastics, on the other hand cannot be reprocessed. However, the problem with thermoplastic reprocessing is that the materials properties are changed if different types of polymers are mixed. Therefore, it is important to keep the types of plastics separated in a recycling program.

To help make it easier to keep plastics separate, the Society of Plastics Industry (SPI) established in 1988 voluntary guidelines to code plastic containers for recycling sorting. SPI recommends that all 8 oz and larger containers, and all 16 oz and larger bottles, have a symbol molded into the bottom of the product indicating the type of plastic it is made from. The symbols established for this purpose are shown in Figure 7-18. For example, code 2 (HDPE) stands for high-density polyethylene and code 3 (V) represents polyvinyl chloride. PETE (1) indicates that the material is polyethylene terephthalate. LDPE (4) is low-density polyethylene, PP (5) is polypropylene, PS (6) is polystyrene, and "other (7)" means any plastic not one of the other codes. These markings provide recyclers a quick method for separating different grades of plastics.

Creating a market for recycled plastics is an effective way to increase recycling. A 258-city survey by the US Conference of Mayors found the biggest barrier to community recycling programs is a shortage of markets for the collected materials. Many plastic bottles, such as laundry detergent bottles, can be made of 100% recycled high-density polyethylene. Plastic products made from recycled plastic can prevent millions of pounds of disposed plastic from reaching landfills each year. As new markets are created for recycled plastics, more programs should be created to collect and recycle plastics.

Figure 7-18.
Resin codes of the Society of Plastics Industry (SPI) help recyclers sort plastic bottles and containers.

PETE   HDPE   V   LDPE   PP   PS   OTHER

SUMMARY

The applications of thermoplastics reach every corner of our daily lives, making the current era the Plastics Age. There are many types of thermoplastics used in industry. Thermoplastics such as polyolefins, styrenic plastics, vinyl plastics, and acrylics are called commodity thermoplastics and find most applications in industry. Engineering thermoplastics tend to compete with metals because of their superior properties. Commodity thermoplastics are used mainly to replace glass, paper, and wood.

Thermoplastics can be formed into various products by extrusion, injection molding, blow molding, rotational molding, thermoforming, and other techniques. Temperatures required to work thermoplastics are much lower than those for metals.

Plastic recycling is important to a clean environment. It is a challenge that must be met by society. Strong efforts are needed to promote plastic recycling so that the amount of solid waste to landfills can be significantly reduced.

IMPORTANT TERMS

Acrylics
Amorphous Polymers
Branched
Copolymer
Creep
Crystalline
Environmental Stress Cracking
Extrusion Blow Molding
Glass Transition
Glass Transition Temperature
Homopolymer
Linear
Molecular Weight
Monomers
Network Polymer
Parison
Polymer
Polymerization
Polyolefin Family
Polypropylene
Polystyrene
Semicrystalline Polymers
Stress Relaxation
Thermoplastics
Thermosets
Vinyl Chloride
Viscoelastic Behavior
Viscoelasticity

QUESTIONS FOR REVIEW AND DISCUSSION

1. What is the major difference between thermoplastics and thermosets?
2. What is a polymer, in terms of molecular structure?
3. How many carbon atoms does an ethylene monomer have? How are they bonded?
4. What are the three basic configurations for polymer molecular chains? How are they related to the property of a polymer?
5. What are the main similarity and difference between extrusion molding and injection molding?
6. How does blow molding differ from injection molding?
7. How is a part formed using rotational molding?
8. In processing polyvinyl chloride, why should the polymer not be overheated?
9. Why is a special incinerator required for disposing of polyvinyl chloride?
10. What is an engineering thermoplastic?
11. Why is polytetrafluoroethylene (PTFE) ideal for use as furniture drawer guides?
12. When subjected to a steady load, what is the major difference in material behavior between steel and thermoplastic?
13. What is creep?
14. What is stress relaxation?
15. What are glass transition and glass transition temperature?
16. The glass transition temperature of polyethylene is –193°F (–125°C). Is this thermoplastic rigid or flexible at room temperature? Why?
17. Why can ultra-high molecular weight high-density polyethylene (UHMW-HDPE) be used as bearings for total joint replacements for human knees and hips?
18. Why is nylon suitable as fiber reinforcement for composites?
19. Why is plastic recycling important?
20. If a ketchup bottle has plastic recycling code 7 molded into the bottom of the bottle, what kind of plastic is the bottle made from?

FURTHER READINGS

1. Ulrich, H. Introduction to Industrial Polymers. München: Hanser Publishers (1992).

2. Berins, M.L. SPI Plastics Engineering Handbook (5th ed.). New York: Van Nostrand Reinhold (1991).

3. Rosen S. L. Fundamental Principles of Polymeric Materials (2nd ed.). New York: John Wiley & Sons (1982).

4. Morton-Jones D. H. Polymer Processing. New York: Chapman and Hall (1989).

5. Kaufman, H. S. and Falcetta J. J. Introduction to Polymer Science and Technology: An SPE Textbook. New York: John Wiley & Sons (1977).

6. Billmeyer Jr., F. W. Textbook of Polymer Science. New York: John Wiley & Sons (1984).

7. Charrier, J.M. Polymeric Materials and Processing. New York: Hanser Publishers (1991).

INTERNET RESOURCES

http://www.4spe.org
The Society for Plastics Engineers

http://www.plasticsindustry.org
Society of the Plastics Industry

http://www.geplastics.com
GE Plastics

http://www.dupont.com/enggpolymers/americas/delrin.html
DuPont Engineering Polymers

http://www.polymers-usa.bayer.com
Bayer Corporation-Polymers

http://www.modplas.com
Modern Plastics

http://www.plasticstechnology.com
Plastics Technology

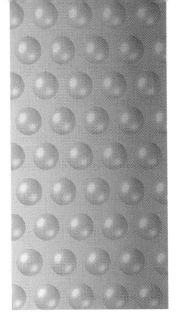

Chapter 8

Thermosetting Plastics

KEY CONCEPTS

Upon completion of this chapter, you should understand:
- ➢ The definition of thermosetting plastics.
- ➢ Polymerization of thermosetting plastics.
- ➢ The classification system used for thermosetting plastics.
- ➢ Properties and characteristics of thermosetting plastics.
- ➢ Processes used to form thermosetting plastics.
- ➢ Applications of thermosetting plastics.

Although plastics have been the fastest growing segment of the United States chemical industry since World War II, their history began long before then. In fact, plastics were first developed in the 1800s. The first commercially successful plastic was developed as a substitute for ivory in the production of billiard balls. This synthetic polymer was called cellulose nitrate. It was available around the middle of the 19th century. The early part of the twentieth century brought the development of phenolic and urea formaldehyde plastics. In 1909, Leo Baekeland introduced these plastics as Bakelite® plastics. These plastics are still used extensively today for a variety of parts and products.

Baekeland's plastics belong to a classification of plastics known as thermosets or thermosetting plastics. *Thermosetting plastics* are synthetic polymers whose molecules cross-link during processing and, therefore, cannot be recycled or reprocessed. They account for approximately 15% of total plastic production. Thermosets include the following families.

- ◆ Phenolics
- ◆ Urethanes
- ◆ Silicones
- ◆ Ureas
- ◆ Epoxies
- ◆ (Amino Resins) Melamines
- ◆ Unsaturated Polyesters

COMPOSITION AND STRUCTURE

As stated in Chapter 7, polymers are based on large molecules that have chain-like, repeating patterns. The basic difference between the thermoplastics and the thermosets is that thermosetting plastics have cross-linked molecular chains, Figure 8-1. This cross-linking creates a strong bond that prevents thermosetting plastics from being reprocessed once they are cured.

POLYMERIZATION

The process of linking molecules together is called polymerization. This was introduced in Chapter 7. There are two basic polymerization reactions—addition and condensation. *Addition polymerization* is essentially a physical linking of one molecule to another. The addition polymerization of polyethylene is illustrated in Figure 8-2. Ethylene (C_2H_4) is a gas consisting of carbon and hydrogen. The carbon atoms have a double bond, as shown by the double line in Figure 8-2. In the polymerization process, the double bond between carbon atoms is broken by a combination of heat, pressure, and catalysts. This process creates single covalent bonds with free radicals at the ends of the monomer. *Free radicals* are atoms having at least one unpaired electron. These single bonds become active allowing other monomers to bond to them in a chain-like fashion. The length of each polymer chain is controlled by the type of catalyst or method of processing. Thermoplastics, discussed in Chapter 7, are formed using addition polymerization.

Thermosetting plastics, on the other hand, are formed through a process known as *condensation polymerization.* Condensation polymerization takes place through chemical reaction rather than by physical reaction. As with addition polymerization, condensation polymerization begins by breaking molecular bonds with heat, pressure, and catalysts. However, unlike addition polymerization, the reaction between the monomers is chemical and a byproduct, usually water, is formed. In Figure 8-3, water (H_2O) is shown as a

Figure 8-1.
Thermosetting plastics have a cross-linked structure.

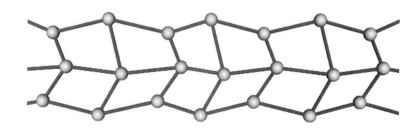

Figure 8-2.
The ethylene monomer consists of two carbon and four hydrogen atoms.

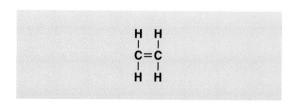

byproduct of the polymerization that takes place in producing polyethyleneglycol terephthalate. This polymer is commonly known by the trade name Dacron®. Terephthalic acid and ethylene glycol react to form Dacron® and produce H_2O. Thermosetting plastics formed by condensation polymerization include the urethanes, phenolics, polyesters, and silicones.

PROPERTIES

Thermosetting plastics have different structures than thermoplastics. Therefore, they have different physical properties. Figure 8-4 compares physical properties of a few popular thermosets and thermoplastics.

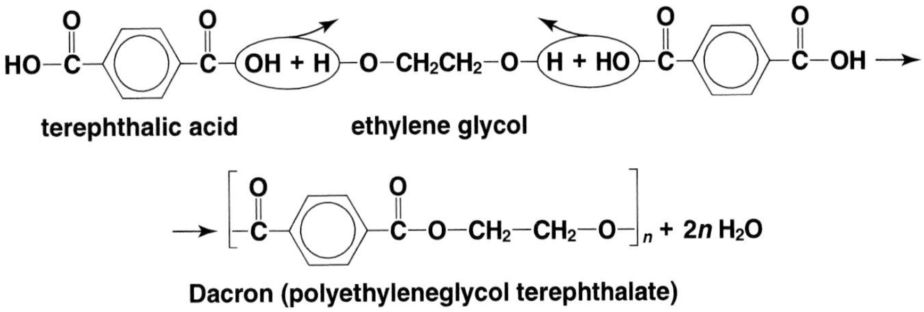

Figure 8-3.
This shows the condensation polymerization formation of Dacron®.

Figure 8-4.
This table lists various properties of selected thermosets and thermoplastics.

PROPERTIES OF PLASTICS					
Plastic Family	Tensile Strength (psi)	Modulus of Elasticity (10^6 psi)	Impact (ft/lb)	Compressive Strength (psi)	Service Temperature
Thermoplastics					
High Density Polyethylene	4,000	0–.35	10	3,000	180°F–250°F
Low Density Polyethylene	1,500	0–.35	No break	3,000	180°F–250°F
Polystyrene	7,500	0–.18	0.3	14,000	140°F–170°F
Polypropylene	4,500		1	7,000	250°F
ABS	6,000		6	8,000	165°F
Thermosetting Plastics					
Epoxy	10,000		0.8	20,000	325°F
Phenolic	7,000	0–2.4	0.40	10,000	300°F
Melamines	10,000		.25–.35		350°F
Urethane	10,000	0–.35			200°F

Generally, thermosets have higher service temperatures, tensile strengths, and compressive strength than thermoplastics. There are differences, however, within each classification as well. The degree of cross-linking in thermosetting plastics determines their properties in many cases.

Although relatively weak when compared to other materials, plastics have the advantage of being low cost, lightweight, corrosion resistant, and excellent insulators. Their specific strength (strength to weight ratio) is very good when compared to materials with higher specific gravity. When reinforced, plastics may exceed metals in tensile strength.

In many plastics applications, stiffness is more important than tensile strength. Think of stiffness as resistance to elastic deformation. Measured by modulus of elasticity (E), the higher the modulus, the lower the elastic deformation. Thermosets are generally somewhat stiffer than thermoplastics.

CLASSIFYING THERMOSETTING PLASTICS

As stated earlier, thermosetting plastics are classified by the following family names.

- Phenolics
- Urethanes
- Silicones
- Ureas
- Epoxides
- Melamines
- Unsaturated Polyesters

Each family has unique characteristics. These make that family especially suited for particular applications. These are discussed in the next sections.

PHENOLICS

The phenolic family of thermosetting plastics are known chemically as phenol formaldehyde. They are formed from a condensation polymerization reaction between phenol and formaldehyde. Phenol is an aromatic molecule. Formaldehyde is an organic compound used as a solvent or preservative. As a group, phenolics are the oldest of the thermosetting plastics and among the least costly. They have high stiffness and impact resistance and are readily moldable. In addition to having high relative strength, phenolics also have good insulating qualities. Although compression and transfer molding are the principal processes used for forming phenolics, they can be extruded and injection molded as well.

There are hundreds of plastic compounds in the phenolic family. Products made from phenolics include various knobs, ashtrays, wheels, radio and TV cabinets, and ignition parts for automobiles, Figure 8-5. The natural color of phenolic plastics is brown or amber and tends to discolor. The compounds are often pigmented with darker colors. Phenolic compounds are often divided into six groups. These are general purpose phenolics, impact resistant grades, electrical grades, heat resistant grades, special purpose grades, and nonbleeding grades.

Figure 8-5.
The phenolic preform on the left is compressed in the die (middle) to form the part on the right.

General Purpose Phenolics

General purpose phenolics are low-cost compounds formulated for noncritical functional requirements. These compounds utilize fillers of "wood flour" and flock. They provide a balance of moderate mechanical and electrical properties.

Impact-Resistant Grades

Impact-resistant grades are higher in cost than general purpose. They are designed to be used in electrical and structural components subject to impact loads. Fillers for this grade are paper, chopped fabric, or glass fibers.

Electrical Grades

Electrical grades have high electrical resistivity. They retain high resitivity under high temperature and high humidity conditions. Mineral fillers are usually used in the electrical grade phenolics.

Heat-Resistant Grades

Heat-resistant grades are designed to retain their mechanical properties in the 375°F to 500°F (190°C to 260°C) temperature range. Some compounds in this grade provide long-term stability at temperatures up to 550°F (288°C).

Special Purpose Grades

Special purpose grades are formulated for exceptional resistance to chemicals or water. They are also formulated for a combination of conditions, such as impact loading and a chemical environment. The chemical-resistant grades are inert to most common solvents and weak acids.

Nonbleeding Grades

Nonbleeding grades are specially formulated for use in container closures and for cosmetic cases. Some plastics have higher diffusion rates or higher permeability. In other words, these plastics have pores which allow liquids or gases to pass. For example, some plastics have plasticizers that bleed-out oily residues, especially at elevated temperatures. This bleed-out is particularly undesirable in food and cosmetic containers. Therefore, specially formulated nonbleeding plastics are used to prevent bleed-out of certain compounds into foods and cosmetics.

URETHANES

Urethanes, or polyurethanes, are a group of plastics based on polyester or polyether resin. Urethanes have excellent tensile strength and elongation, good ozone resistance, and good abrasion resistance. Certain combinations of hardness and elasticity can be achieved with urethanes that cannot be obtained in other plastics. Urethanes are fairly resistant to many chemicals, certain fuels, and oils.

Rigid polyurethane plastics are used as solid tires on heavy equipment and for rollers on material handling and printing equipment. They are also used for shock impact devices and automobile bumpers. Rigid urethanes have densities ranging from about 5 lb/ft^3 to 50 lb/ft^3.

Urethane foams are made by adding a carbon dioxide-producing compound or by reacting diisocyanate with an active hydrogen-containing compound. Urethane foams are used for insulating refrigerated units, padded areas of automobiles, sponges, cavity fillings for boats, furniture cushions, as well as many other applications. Flexible urethane foams have densities ranging from about 1 lb/ft^3 to 5 lb/ft^3.

SILICONES

Unlike most other plastics, silicones are not hydrocarbons. They are composed of monomers with oxygen atoms attached to silicon atoms. The polymers in silicones are bonded to each other by silicon-oxygen-silicon cross-linking, Figure 8-6.

Silicones have low moisture absorption. They also resist attack by petroleum products and acids. Silicones are considered inorganic plastics and have higher heat resistance than organic plastics.

Figure 8-6.
This shows the silicone polymer structure.

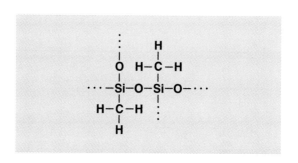

Silicones are considered premium plastics which contributes to their high cost and limited use. Silicones are often used in high performance products in the aircraft and aerospace industries. Silicones are also used in the electronics industry and as sealants and gaskets for numerous applications.

EPOXIES

Commonly known for their use as adhesives, epoxies are premium thermosets with excellent relative mechanical and electrical properties. Although more costly than many plastics, epoxies are invaluable when used as chemical-resistant coatings and for encapsulating electronic components.

Epoxies are heat resistant, hard, and achieve tensile strengths as high as 12,000 lbs/in^2 (82 MPa). They have very high adhesion to metals and nonmetals. Epoxies also have dielectric strength to 500 volts/mile (22 volts/meter). Epoxies cast easily with little shrinkage and are cured into a finished plastic by a catalyst or with hardeners containing active hydrogen.

MELAMINES AND UREA

The melamine and urea plastics represent two large groups of polymer plastics often placed in the broader classification *amino resins,* or *aminos.* The polymer plastics produced from amino resins are usually the hardest of all plastics. As a class, they do not have good impact strength. Amino resins are manufactured by reacting either melamine or urea with formaldehyde. They are clear or white resins which can easily be colored.

Supplied as molding powders or granules, melamine and urea are used for a variety of applications. Molded products of urea include electrical switch plates, wiring devices, housings for small appliances, and buttons. Melamines may be best known for dishes and tableware. However, they are also used for tabletops, countertops, and as an adhesive in laminated wood products such as plywood.

UNSATURATED POLYESTERS

Unsaturated polyesters represent a large group of synthetic resins produced by condensation of acids with an alcohol or glycol. The acid may be maleic, phthalic, or itaconic. The alcohol or glycol may be allyl alcohol or ethylene glycol. When polymerized, cross-linking occurs between molecules. Thermosetting polyester resins are commonly used as a binder for glass-fiber reinforced plastic products. Unsaturated polyester resins have high strength, high adhesion, and good chemical resistance.

Some polyester resins are used to produce textile fibers and sheet or film stock. Approximately 75% of all polyester production is used in glass-reinforced plastics as moldings or laminates. Unreinforced polyesters have limited use, primarily as casting resins in both flexible and rigid forms. Cast polyester products include buttons, electrical components, and architectural shapes.

PROCESSING

Generally, fewer processes are used in forming thermosetting plastic than are used for thermoplastics. This is due to the basic structure difference between the plastics. Since thermosets are composed of cross-lined polymers, they cannot be remelted for recycling. This generally means less flexibility in choosing a manufacturing process. The most commonly used manufacturing processes for thermosetting plastics are compression molding, transfer molding, casting, and reaction injection molding (RIM).

COMPRESSION MOLDING

In *compression molding,* powder or granule thermosetting plastic are placed in a die. Pressure is added and the powder forms solid plastic, Figure 8-7. The raw plastic is often phenolic. However, other thermosets may be used as well. The process usually takes from 3 to 20 minutes, depending on the type of plastic, size of part, and press specifications.

Figure 8-7.
These schematics show three examples of matched dies for compression molding.

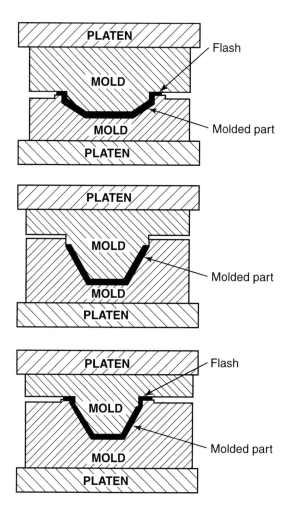

Mold temperatures range from 280°F to 400°F (140°C to 240°C). Pressures in the range of 1,000 psi to 10,000 psi are applied by hydraulic presses, Figure 8-8. The press can be a small hand-operated bench model or a large automated floor model. Molds are usually highly polished tool steel of single- or multi-cavity design.

The heat and pressure cause the thermosetting plastic to melt and flow into the cavity openings. When polymerization is completed, the die is opened and the part removed. As the polymerization of thermosetting plastics causes cross-linking to occur, the dies do not need to be cooled, as is the case with thermoplastics.

TRANSFER MOLDING

Transfer molding is basically the same as compression molding, except that a double chamber is used in the die design. This minimizes direct pressure on the die cavity. Transfer molding is used where the parts have thin sections or intricate detail. Figure 8-9 compares a cross section of a die used for compression molding with a die used for transfer molding.

Figure 8-8.
A—In compression molding, the preform is placed in the mold. The pressure of the mold closing creates the final part. B—In transfer molding, the mold is closed. The preform is placed in a reservoir. A plunger compresses the preform and forces it into the mold to create the final part.

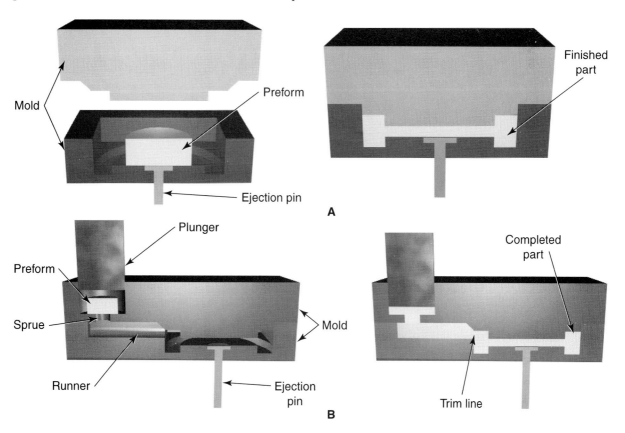

In compression molding, the pressure is applied directly on the plastic and cavity. In transfer molding a thermoset plastic disk called a *preform* is placed in the top half of the die and heated. The preform is made to the approximate shape and size of the die cavity. Pressure melts the preform and forces it through a system of runners into the mold cavity.

The temperatures, pressures, and equipment used in transfer molding are the same as those used for compression molding. One advantage of transfer molding is that little post processing is required. Very little flash is formed. *Flash* is waste around the parting line of the die halves.

REACTION INJECTION MOLDING (RIM)

Reaction injection molding (RIM) is typically used in forming products from polyurethane foams, Figure 8-10. During this process, polymer reactants are pumped from separate reservoirs into a mixing chamber, then into a heated mold. The mold is where the product takes shape and is cured. Large automobile parts, such as fenders and dashboards, are made using this process. Architectural trim and molding is often made using this process as well.

Figure 8-9.
This hydraulic press is used for compression molding. (Hull-Finmac, Inc.)

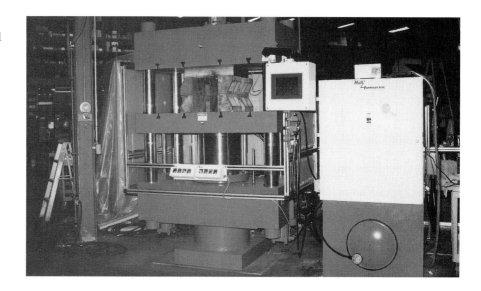

Figure 8-10.
In reaction injection molding (RIM), two polymer components are mixed and injected into a mold. The mold is held together under pressure as the plastic cures.

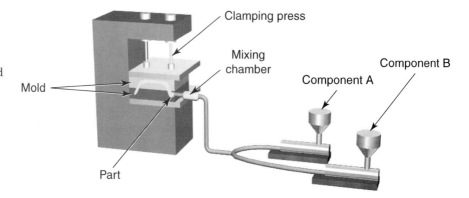

SUMMARY

Thermosetting plastics account for approximately 15% of all polymer plastic production. Thermosetting plastics differ from thermoplastics in their structure. Thermoplastics are formed by addition polymerization. Thermosetting plastics are formed by condensation polymerization. This creates cross-links across polymer chains. The cross-linking forms strong bonds which are not easily broken. Therefore, thermosetting plastics, unlike thermoplastics, cannot be remelted and used again after initial processing.

Thermosetting plastics are generally harder and more brittle than thermoplastics. Thermosetting plastics also generally have greater thermal stability than thermoplastics. However, they are more difficult to form. Thermosetting plastics are used in a variety of products, including adhesives, cookware, sealants, insulation, electrical components, and laminated wood products.

IMPORTANT TERMS

Addition Polymerization
Amino Resins
Aminos
Compression Molding
Condensation Polymerization
Electrical Grades
Flash
Free Radicals
General Purpose Phenolics
Heat-Resistant Grades
Impact-Resistant Grades
Nonbleeding Grades
Preform
Reaction Injection Molding (RIM)
Special Purpose Grades
Thermosetting Plastics
Transfer Molding

QUESTIONS FOR REVIEW AND DISCUSSION

1. What is the basic structural difference between thermosetting plastics and thermoplastics?
2. What contributions did Leo Baekeland make to the development of plastics?
3. Explain the process that occurs during condensation polymerization.
4. What controls the length of each polymer chain during production?

5. In general, how do thermosetting plastics compare to thermoplastics in mechanical and physical properties?

6. What are the different groups of phenolics?

7. What special characteristics do urethane plastics have?

8. What are three of the principal uses for urethane plastics?

9. How do silicones differ from other plastics in composition?

10. What plastic families are classified as aminos?

11. What application represents the greatest use of polyester production?

12. What is the basic difference between compression molding and transfer molding?

13. How does reaction injection molding (RIM) differ from the injection molding process for thermoplastics?

FURTHER READING

1. Berins, M. L. SPI Plastics Engineering Handbook (5th ed.). New York: Van Nostrand Reinhold (1991).

2. Brady, George S. and Clauser, Henry R. Materials Handbook (13th ed.). New York: McGraw-Hill, Inc. (1991).

3. Kaufman, H. S. and Falcetta, J. J. Introduction to Polymer Science and Technology: A SPE Textbook. New York: John Wiley and Sons (1977).

INTERNET RESOURCES

http://www.4spe.org
Society of Plastics Engineers

http://www.plasticsindustry.org
Society of the Plastics Industry

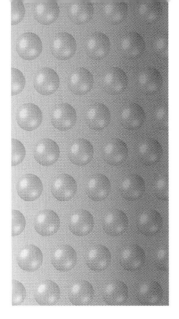

Chapter 9

Wood and Engineered Wood Products

KEY CONCEPTS

Upon completion of this chapter, you should understand:

➤ The composition and structure of softwoods and hardwoods.
➤ The advantages of wood as an engineering material.
➤ The processes used for harvesting, seasoning, and grading lumber.
➤ The relationship of moisture content and seasoning to the properties and stability of wood.
➤ Conservation and reforestation.
➤ Structural composite lumber.
➤ Structural panel products.

Wood has served the needs of humans and extended human potential for thousands of years. For centuries it was the material of choice for shelter, fuel, transportation, recreation, and manufacturing. Although metals, plastics, and composites have replaced wood for many applications, it remains a popular choice for many others.

While synthetic materials, metals, and ceramics are often less expensive or stronger, wood has many advantages of its own. The following is a partial list of the advantages of wood.

♦ Abundant
♦ Renewable
♦ Easily shaped
♦ Aesthetically pleasing
♦ Strong
♦ Noncorrosive

Although many of the large mature trees and forests have been depleted, scientists have developed methods of replenishing forests with species of fast growing trees. Wood remains one of the warmest, most aesthetically pleasing materials for construction, manufacturing, and architectural applications. It is so aesthetically pleasing that many synthetic materials used for furniture and architectural applications are produced to simulate wood.

COMPOSITION AND STRUCTURE

Although most commonly classified as a natural polymer, wood also fits the definition of a composite. Wood consists of long cellulose fibers held together by lignin. It is actually a composition of four things: cellulose, hemicellulose, lignin, and extractives.

Cellulose ($C_6H_{10}O_5$), a naturally occurring thermoplastic polymer, represents approximately 40–50% of the total composition of wood. Cellulose is manufactured from units of glucose formed by photosynthesis.

Hemicellulose is a low-molecular-weight polymer formed from glucose. It makes up approximately 12–28% of wood. Usually, hemicellulose is a branch chained polymer rather than linear, Figure 9-1.

Lignin is a hard, amorphous polymer. It comprises about 12–28% of wood and binds the cellulose and hemicellulose together. Hardwoods usually have less lignin and more hemicellulose than softwoods. This contributes to the higher density of hardwoods.

Extractives include oils, silica, and other trace elements. They are a small percentage, approximately 5%, of the total composition.

MICROSTRUCTURES

At the microscopic or *microstructural level,* trees are composed of elongated cells made of cellulose, hemicellulose, lignin, and extractives. Each cell consists of long fibers of cellulose and hemicellulose bound together by lignin, Figure 9-2. The fibers have an aspect ratio of 100 or more. The aspect ratio is a ratio of length to diameter. In other words, the fibers are 100 times as long as they are wide. Each cell is made up of several layers built up from microfibrils. A *microfibril* is a submicroscopic, elongated bundle of cellulose comprising the cell wall of a plant or tree.

MACROSTRUCTURES

The microscopic cells, fibers, and microfibrils link together to form what you see as a tree. This larger view is called the *macrostructure* of a tree. Figure 9-3 shows a cross section of a tree to illustrate the various parts of the macrostructure.

At the very center of the tree is a dark, soft, sponge-like material known as *pith.* Pith stores food and transfers it to other layers of the tree.

Figure 9-1.
A—A representation of polymer chains in a linear polymer. B—A representation of polymer chains in a branched polymer.

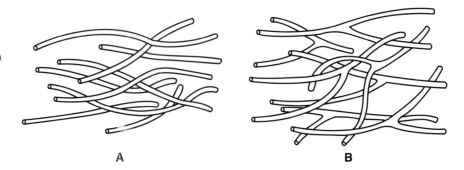

A   B

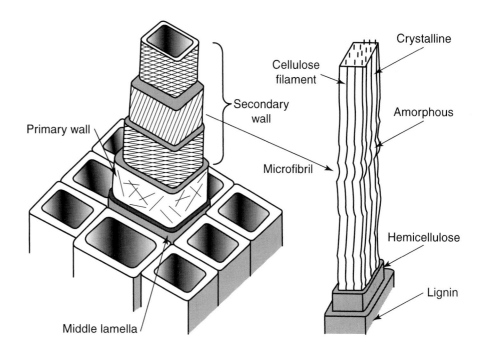

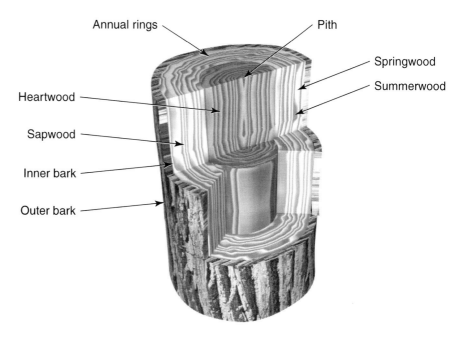

Surrounding the pith is the *heartwood.* The heartwood is the older growth of the tree. The cells in the heartwood are no longer living. However, it does not decay and provides much of the structural strength to the tree.

Next to the heartwood is the *sapwood.* Sapwood is lighter in color than the heartwood. It consists of living cells that transport water to the leaves. As the tree grows, the inner cells of the sapwood change to heartwood.

The thin layer of cells next to sapwood is the *cambium layer.* The cambium layer is the newest growth on the tree.

The *outer bark* protects the living cells of the tree. It helps the tree retain moisture, insulates against heat and cold, and protects the tree from insects.

Between the outer bark and the cambium layers is the *inner bark.* The inner bark, or *phloem,* is the food pipeline for the tree. The inner bark lives for only a short period of time and is then converted to outer bark. New inner bark is then formed by the cambium layer.

While most cells grow vertically, there are groups of cells called *medullary rays* that radiate from the pith and extend to the outer bark. These radial cell groups transfer and store food.

Looking at a cross section of a log or tree trunk reveals distinctive rings called *annual rings,* Figure 9-4. Each annual ring represents one year's growth. Therefore, by counting the number of rings, the age of the tree can be determined. Each annual ring consists of a wider band of lighter colored cells and a narrow band of darker colored cells. The wider band is called the *early wood* or *springwood.* The springwood represents the growth occurring in the spring. The lighter appearance of this band is a result of the cell walls being farther apart due to availability of moisture and absorption. The narrow band is called the *late wood* or *summerwood.* Since there is less moisture in the summer, the cells produced during that time of year are more dense and cell cavities smaller. Thus, the darker appearance and narrower band.

HARVESTING AND PROCESSING

Once a tree is harvested, it is processed into usable lumber. Trees are a renewable resource. However, it is important to make sure that the resource is renewed so it does not disappear. Conservation and reforestation are fields dedicated to maintaining a continued supply of trees.

Figure 9-4.
Annual rings, heartwood, sapwood, inner bark, and outer bark are all important features of a tree.

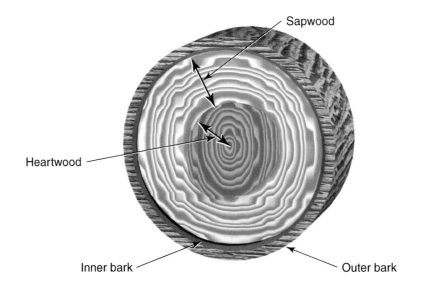

The lumber used to build homes, manufacture furniture, produce paper, and produce plastic comes from many different species of trees. However, all trees are classified as either a softwood or a hardwood.

SOFTWOOD

Softwood lumber production is about 75% of all lumber production. Softwoods are used primarily for construction. However, they are also used for certain furniture styles and in the production of paper and sheet products. Flakeboard and plywood are examples of sheet products.

Softwood trees are *coniferous* species. These tree species bear cones and retain their leaves year–round. Softwood trees are harvested from forests and tree farms in three major regions of the United States, Figure 9-5. Botanically, softwoods are gymnosperms or conifers. Their seeds are not enclosed in the ovary of the tree's flower. Anatomically, softwoods are non-porous and do not contain vessels.

The *Southern Pine Region* is located in the southern half of the country. It stretches from the Atlantic Ocean to the Gulf of Mexico. The softwood species harvested from this region include shortleaf pine, longleaf pine, loblolly pine, slash pine, and cypress.

The *Western Wood Region* encompasses the western third of the country. It is the largest geographic region of softwoods and stretches from southwestern Texas to northwestern Washington. Species commonly harvested in this region are Douglas fir, ponderosa pine, red cedar, hemlocks, white fir, Sitka spruce, western larch, lodgepole pine, and sugar pine. Within the Western Wood Region, in northern California, is the *Redwood Region.* This region produces redwood and Douglas fir.

HARDWOOD

Hardwood lumber production is about 25% of the total wood production. The primary use of hardwoods is for furniture. However, hardwoods are used in some construction applications, such as floors and trim. In fact, hardwood floors have long been a very desirable feature in houses.

Hardwood lumber is harvested from deciduous trees. *Deciduous* trees are broadleaf species that lose their leaves each year. Botanically, hardwoods are angiosperms. Their seeds are enclosed in the ovary of the tree's flower. Anatomically, hardwoods are porous and contain vessels. A vessel is a wood cell with open ends.

Most hardwoods are harvested from trees found in the eastern half of the United States, Figure 9-6. These species are located in the Northern Forest and Central and Southern Hardwood Forest regions. The *Northern Forest Region* extends from Maine to Wisconsin. This region produces maple, birch, beech, oak, and black cherry. The *Central and Southern Forest Region* stretches from the East Coast west to beyond the Mississippi River, and from the Great Lakes south to the Gulf of Mexico. It produces most of the oak, walnut, hickory, yellow poplar, gum, and basswood harvested.

Figure 9-5.
This maps shows the softwood tree regions of the United States.

Key	Region	Species	Approximate Yearly Production	Lumber Association	Grading Authority
■	Western Wood Region	Douglas fir Ponderosa pine Western red cedar Incense cedar Western hemlock White fir Engelmann spruce Sitka spruce Western larch Lodgepole pine Idaho white pine Sugar pine	16.9 billion board feet	Western Wood Products Association	Western Wood Products Association West Coast Lumber Inspection Bureau
□	Redwood	Redwood Douglas fir	2.7 billion board feet	California Redwood Association	Redwood Inspection Service
▨	Southern Pine Region	Longleaf pine Shortleaf pine	6.7 billion board feet	Southern Forest Products Association	Southern Pine Inspection Bureau

Figure 9-6.
This maps shows the hardwood tree regions of the United States.

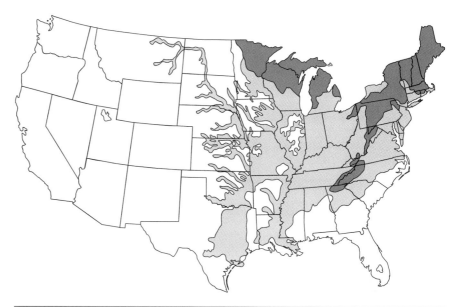

Key	Region	Hardwoods
■ (dark)	Northern Forests	Maple Birch Beech Ash Black cherry
□ (light)	Central and Southern Hardwood Forests	Oaks Yellow poplar Gum Hickory Black walnut Basswood

FROM FOREST TO MILL

Once selected and cut down, the tree is called a *log.* Logs are transported from the forest to the sawmill by truck, railroad, or river. The *sawmill,* or mill, is where logs are turned into usable lumber. At the mill, logs may be stored in ponds until ready for processing.

Processing logs begins with debarking. *Debarking* is removing the bark from the log. This is done by peeling, scraping, or forcing it off with high-pressure water jets. The debarking process reveals the irregularities, such as knots. The sawyer then determines the best way to saw the log to avoid waste.

The debarked log is placed on a carriage that feeds the log into a head saw, Figure 9-7. The log is then rough sawn into timbers, planks, and boards. Rough sawn boards are cut to approximate size and shape. They do not have finished, smooth surfaces. There are two primary methods for sawing logs into boards—plain-sawing and quartersawing.

Figure 9-7.
A head saw is used to saw logs into rough cut boards. As the carriage moves along the length of the log, the saw cuts the log.

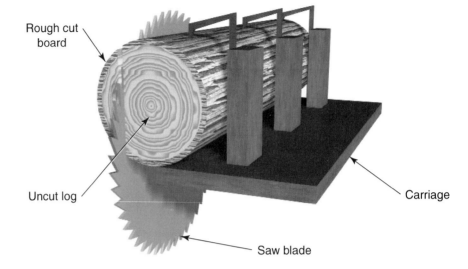

Figure 9-8.
Lumber is generally cut in one of two ways.
A—Plain-sawn.
B—Quartersawn.

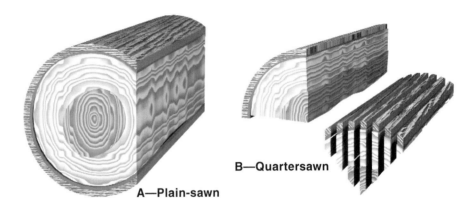

Plain-sawing is done by making cuts tangent to the annual rings. This produces the grain pattern shown in Figure 9-8. The main advantage of plain-sawing is it produces more lumber from a log than quartersawing. The disadvantage is the resulting lumber may warp and shrink more than quartersawn lumber.

Quartersawing is done by making cuts radially to the annual rings. This produces a straight grain pattern. The advantages of quartersawn lumber are less warp and shrinkage, better wear, and reduced surface checking. The main disadvantage is that there is more waste.

CONSERVATION AND REFORESTATION

Historically, the logging industry has been an "extractive industry." During the nineteenth century, only about 20 percent of the tree was used. The rest was left behind on the forest floor. As the forests were used up, the logging industry simply moved farther west to new forests.

In the early part of the twentieth century, the logging industry began to change from "extractive" to "harvesting." As forests disappeared with no new forests to move on to, conservation and reforestation became important. *Conservation* is a program of preserving and protecting natural resources. *Reforestation* is the planting of new trees to replace those cut down by logging. Also, efforts were made to use more of the tree and waste less.

The Weyerhauser Corporation helped lead the movement by initiating the *Tree Farm Certification* system in 1941. This system supported reforestation through genetic engineered trees and effective forest management. Today, wise forest management and genetic engineering advances help ensure that trees as a natural resource will be around for future lumber needs.

SEASONING LUMBER

Freshly-cut timber has an extremely high moisture content. Wood that has not been dried is called *green*. A green tree contains as much as 200% or higher moisture content over dried wood. Much of this moisture is in the form of free water within the wood's cell cavities, Figure 9-9. This high moisture content makes wood unsuitable for use. Wood must be seasoned to be suitable for use. *Seasoning* is the process used to reduce the moisture content of wood from its green state to between 4% and 19%.

Reducing Moisture Content (Seasoning)

The first stage in reducing moisture content is removing free water from cell cavities. After free water is removed, the remaining moisture is in the cell walls, Figure 9-9. This point is referred to as *fiber saturation point.*

Figure 9-9.
A typical wood cell.
(Forest Products Laboratory,
USDA Forest Service)

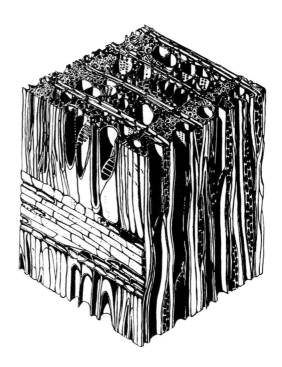

The moisture content at fiber saturation point is approximately 30%. The next step in the process is to reduce the moisture in the cell walls. This results in the final moisture content between 4% and 19%.

The exact moisture content depends on the type of use and geographic location. Lumber used for construction is usually dried or seasoned to a moisture content between 12% and 19%. Cabinet grade lumber is seasoned to between 4% and 12%. The exact moisture content within these ranges is determined by geographic location and use. Wood is hygroscopic. *Hygroscopic* means that the wood readily absorbs and retains moisture. In higher humidity climates, wood absorbs moisture, thus increasing the moisture content. In low humidity climates, wood may lose moisture, thus reducing the moisture content. Therefore, the humidity level in the geographic area often dictates what the final moisture content of the lumber should be.

Effect of Changing Moisture Content

Moisture content affects the physical, mechanical, and dimensional properties of wood. Once a tree is cut into lumber, moisture evaporates over a period of time until it reaches what is known as the *equilibrium moisture content.* At this point, the moisture content in the air and the moisture content in the wood are equal. As the moisture content in the air rises or falls, so does the moisture content in the wood. This change causes shrinking and swelling of the wood. Sealing the wood with a finish can reduce the instability caused by changes in humidity. Plain-sawn lumber generally undergoes greater dimensional and shape changes than quartersawn lumber. Also, hardwoods usually shrink more than softwoods.

Shrinkage

Shrinkage can easily be seen in everyday life. For example, wood for decks usually has very high moisture content. When installed, the floorboards are often placed side by side with no space in between. Within a few days, the sun and wind cause evaporation of moisture in the wood. This causes the boards to shrink and, therefore, create gaps between each board. In this same application, shrinkage of length cannot be seen.

In plain-sawn lumber, the greatest shrinkage occurs tangent to the annual rings, Figure 9-10. The next greatest shrinkage occurs in the radial direction. Shrinkage tangent to the annual rings may be 50% to 67% greater than in the radial direction. The reason shrinkage occurs more tangent to the annual rings is due to a greater loss of moisture in the cell cavities causing the cell walls to move closer together. The smallest shrinkage occurs in length.

Warpage

Wood is an unstable material. It can change dimensions and shape as the humidity in the air varies. As wood loses and gains moisture, it can twist and bend in various directions. This distortion is called *warp.* The distortion can take one of several shapes or forms, Figure 9-11.

Figure 9-10.
Lumber shrinks when it is dried. Notice how more shrinkage occurs in the tangential direction.

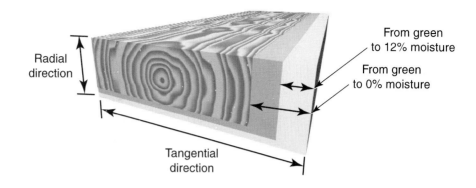

Figure 9-11.
Warp can occur in many forms.

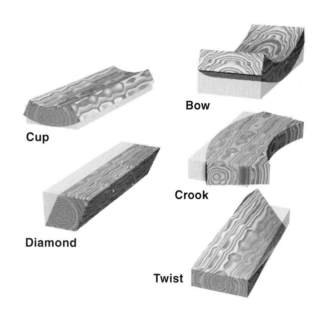

Mechanical Properties

Moisture content affects the mechanical properties, or strength, of wood. As wood dries, the fibers in the cell walls stiffen, increasing strength. For example, take a green branch and a dried branch. Compare the strength of each. A green branch bends easily. On the other hand, a dried branch is much stiffer and resists bending. The increase in strength is a result of increased density of wood fibers as they dry and stiffening of fibers as moisture is lost. The density may increase by as much as 20% from the green state to seasoned wood.

Physical Properties

The physical properties affected by moisture content include electrical resistance and decay resistance. At a moisture content below 20%, wood easily resists decay. Above 25%, decay can easily begin.

As moisture content decreases in wood, the electrical resistance increases. The drier the wood, the lower the electrical conductivity and the greater the electrical resistance. This principle is the basis for instruments used to determine moisture content in wood.

DETERMINING MOISTURE CONTENT IN WOOD

The moisture content in lumber is determined in one of two ways. One way is to use a moisture content meter. The other way is to use a calculation called the oven dried method.

A moisture content meter measures the electrical resistance of the wood. Two probes are inserted into the wood. The measured resistance is converted to a moisture content percent. Although these meters are generally accurate to within 1%, certain conditions may affect the reading. High humidity, the location of board being tested, and fog can all affect the reading taken with a meter.

The second method is often called the *oven dried method.* In this method, samples approximately 1" long are selected from the lumber. The samples are carefully weighed and the weight is recorded. The samples then are placed in an oven that has been preheated to 215°F. After two hours, the samples are removed and weighed. The weight is recorded and the samples are placed in the oven for an additional 15 minutes. The samples are weighed again and the weight recorded. This process continues until the weight no longer changes. The green weight and the final oven dry weight are then used in the following formula.

$$\text{M.C. \%} = \frac{\text{green wt. - oven dry wt.}}{\text{oven dry wt.}} \times 100\%$$

For example, the green weight of a sample is 2.50 oz. The final oven dry weight is 1.55 oz. The moisture content is calculated as follows.

$$\text{M.C. \%} = \frac{2.50 - 1.55}{1.55} \times 100\% = \frac{.95}{1.55} \times 100\% = .6129 \times 100\% = 61.29\%$$

GRADING STANDARDS

Quality standards are used to grade lumber based on use. The standards for grading lumber are prepared by region. However, all standards conform to the American Softwood Lumber Standard 20 - 70 and standards established by the National Hardwood Lumber Association.

SOFTWOOD GRADES

Softwood lumber is classified as *Yard Lumber* and *Factory and Shop.* Yard Lumber is subclassified as *Boards, Dimension,* and *Timber,* Figure 9-12. Dimension lumber and Timbers are used for structural applications in construction. Factory and Shop lumber is for special uses, such as windows and doors. Within these classifications, softwood lumber is graded as either select or common grades.

Select grades have a good appearance and may be used where a natural finish is required. Siding, paneling, and flooring are generally manufactured from select grade softwoods. The two highest grades of select are *A Select* and *B Select.* These two are combined for most species into a grade known as *B and Better.* Other grades with lower quality include *C Select* and *D Select.* The higher the letter, with A the highest, the better the grade.

Figure 9-12.
This is a common classification system for softwood lumber.

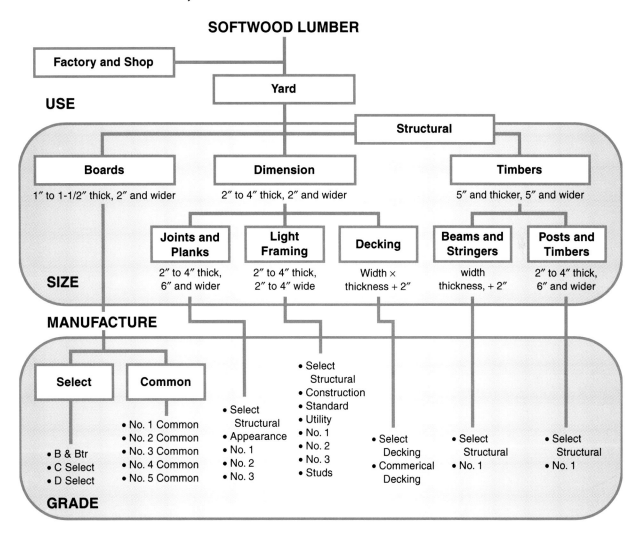

Common grades are usually not suitable for trim, molding, paneling, or other finishing products. These grades are used for general construction and utility applications.

Light framing lumber is lumber up to 4″ thick and wide. The grades for light framing are *construction, standard,* and *utility.* These grades are used in applications where high strength is not a requirement, such as studs, plates, and blocking. *Select structural* No. 1, No. 2, and No. 3 are also light framing grades. These grades are designed for applications where higher strength is required, such as roof trusses.

Timbers are structural members 5″ and thicker or 5″ and wider. Timbers are used as posts and beams. Commonly used grades of timbers are *select structural* and *No.1.*

HARDWOOD GRADES

Hardwood grades are based primarily on visual appearance and size. The highest quality of hardwood is *Firsts and Seconds (FAS).* FAS grade hardwood must be 6" or wider and at least 8' long. It must also be 85% to 90% clear on the poorest face. The term *clear* means free of defects and discoloration. A clear face has no knots, cracks, or color blemishes.

The next lower grade is Select. Hardwood *Select grades* have similar size requirements to FAS. However, the best face must be 85% to 90% clear.

Common grades are lower quality than Select. They are 3" to 4" wide, 2' to 3' long, and 2/3 (66.67%) clear on the best face.

PROPERTIES OF WOOD

Wood is a natural polymer. Therefore, it exhibits properties common to most polymers, both natural and synthetic. It does not have strong atomic bonds, unlike metals and ceramics. Wood is also a nonconductor of both heat and electrical energy. The density of wood can vary greatly from one species to another.

The mechanical properties of wood are directly related to its physical properties. Moisture content, density, and specific gravity all affect the mechanical strength of wood. Using research conducted by the United States Forest Product Laboratory, predictions of strength can be made using the specific gravity of individual species.

DENSITY AND SPECIFIC GRAVITY

Although specific gravity and density are related, they do not have identical meanings. *Density* is the mass or weight per unit of volume and calculated:

$$\text{density} = \frac{\text{weight of wood with moisture}}{\text{volume of wood with moisture}}$$

The density is different by species and family or classification. Softwoods contain less of the higher density summer or latewood. Therefore, they are generally less dense and have lower strength than hardwoods.

Specific gravity is the ratio of density of a material to the density of water at a standard temperature. The formula for calculating specific gravity in wood is:

$$G_b = (W_d/V_d)/P_w$$

where G_b is specific gravity of wood, W_d is the oven dry weight of wood, V_d is the volume of dry wood, and P_w is the density of water.

MECHANICAL PROPERTIES

The mechanical properties of wood are highly *anisotropic.* This means that they differ greatly based on direction. Wood has higher strength with the grain (longitudinal direction) than across the grain, Figure 9-13. Imperfections that may be found in wood include knots, knot holes, and checks. These imperfections may further reduce the mechanical strength.

Wood free of imperfections can have longitudinal tensile strength in the range of 10,000 psi to 20,000 psi. Wood with imperfections may have tensile strength below 5000 psi.

A mechanical property important to many structures manufactured of wood is bending strength. ***Bending strength*** is a measure of the stiffness of the wood or the ability of the wooden member to resist loads applied perpendicular to the longitudinal axis. When a force is applied in this direction, two types of stress are exerted on the member. A compression stress is applied to the top edge of the members and a tensile stress to the bottom edge, Figure 9-14. This property is particularly important for spanning structural members. Beams, ceilings joists, and floor joists are a few examples of this. The measures of bending strength are modulus of elasticity and modulus of rupture.

Modulus of Elasticity

The modulus of elasticity for woods ranges from 500,000 to 2,800,000. The higher the MOE the higher the resistance to bending stresses. Figure 9-15 shows modulus of elasticity for some popular woods. Modulus of elasticity is calculated using the formula on the following page.

Figure 9-13.
The anisotropic properties of wood means it has more strength across the grain.

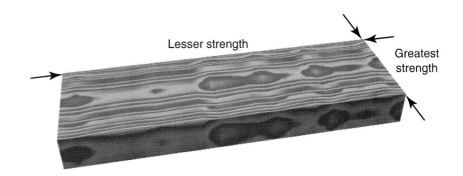

Figure 9-14.
There are different compression and tension stresses on a wood beam in a bending strength test.

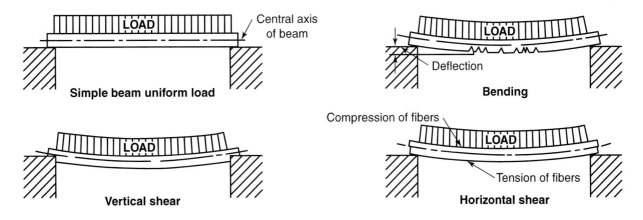

MODULUS OF ELASTICITY FOR WOOD	
Species	**Modulus of Elasticity (psi)**
Softwood	
Pine	1,100,000
Cedar	1,200,000
Douglas Fir	1,800,000
Spruce	1,300,000
Redwood	1,400,000
Hardwood	
Oak	1,800,000
Maple	1,500,000
Birch	2,000,000
Ash	1,600,000

$$MOE = PL^3/48\ ID(psi)$$

where P is the load in pounds, D is the deflection in inches at the midpoint of the span with a load applied, L is the length of the span (inches), and I is the moment of inertia. The moment of inertia is a function of the beam size and calculated:

$$I = Width \times Depth^3/12''$$

Figure 9-16 shows the standard ASTM procedure for testing bending strength. A standard 2″ x 2″ x 30″ specimen is placed on a frame and a force is applied at the midpoint of the span. The load and the deflection from the initial plane are recorded and used to calculate the modulus of elasticity.

Example: A standard 2″ x 2″ x 30″ specimen of white oak is placed in the testing device. The test specimen is placed between the vertical supports so that one inch is resting on each support member. A force of 1500 lbs. is applied at the midpoint of the remaining 28″ span and the deflection at midpoint is measured as .252″. The modulus of elasticity is calculated as follows.

$$MOE = PL^3/48\ ID = 1500 \times 28^3/48\ (2 \times 2^3/12) \times .252 =$$
$$32,928,000/48\ (1.333)\ .252 = 32,928,000/16.128 = 2,041,666.67\ psi$$

Modulus of Rupture

The second measure associated with bending strength is the *modulus of rupture.* Modulus of rupture is a measure of a material's resistance to failure when a load or force is applied perpendicular to its horizontal axis. The testing procedure for modulus of rupture is similar to that for modulus of elasticity. The specimen and equipment are the same. The difference is that the force is applied until fracture occurs. Modulus of rupture is calculated using the following formula.

$$MOR = 1.5PL/BD^2$$

where P is the force or breaking load, L is the span, B is the width, and D is the depth.

Example: A standard 2″ x 2″ x 30″ specimen of white oak is placed in the testing device and loaded to failure. The span is 28″ and the breaking load is recorded as 2900 lbs. The modulus of rupture (MOR) is calculated:

$$MOR = 1.5PL/BD^2 = 1.5(2900)28/ \ 2 \times 2^2 = 121,800/8 = 15,225 \ psi$$

TENSILE AND COMPRESSION STRENGTH

In most applications, tensile strength is not an important factor for wood. However, there is a standard ASTM test for determining tensile strength. The specimen for tensile testing is shown in Figure 9-17.

Figure 9-16.
An ASTM bending strength test being performed on a wood sample.

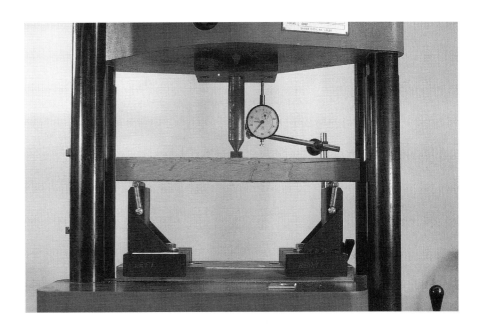

Figure 9-17.
This wood specimen is prepared for a tensile test.

On the other hand, compression strength in wood is important in most applications. It is normally tested both parallel and perpendicular to the grain. This is because wood is often used both horizontally and vertically. Compression strength is calculated by dividing the maximum load by the cross-sectional area of the specimen. The standard specimen dimensions are 2″ x 2″ x 8″.

$$\text{Compressive Strength} = \frac{\text{Breaking Load}}{\text{Cross-Sectional Area}}$$

Figure 9-18 shows tensile and compression strengths of some common softwoods and hardwoods in the longitudinal and radial directions. The significant difference in strength between the two directions of testing are the result of wood's anisotropic nature.

PHYSICAL PROPERTIES

Dry wood is a relatively stable material. It does not expand or contract to any significant degree with temperature changes.

Wood does not conduct energy. Therefore, it acts as an insulator against both thermal and electrical energy. This property also allows wooden structural members to retain much of their strength when charred by fire. Heavier structural components have a tendency to char only on the outside when in a fire. This characteristic is known as *ablation.* Ablation helps protect the wood against further burning.

Figure 9-18.
This table shows tensile and compression strengths of various wood species.

TENSILE AND COMPRESSION STRENGTHS OF WOOD			
Species	Compression Strength Parallel to Grain (psi)	Compression Strength Perpendicular to Grain (psi)	Fiber Stress Proportional Limit (psi)
Softwoods			
Douglas Fir	3784	382	7,400
Western Hemlock	3756	399	3,400
Loblolly Pine	3511	389	4,100
Longleaf Pine	4321	479	5,200
Shortleaf Pine	3527	353	3,900
Cedar (Western Red)	2774	244	53,000
Redwood	4210	424	4,800
Spruce (Sitka)	2670	279	3,300
Hardwoods			
Birch	3380	428	4,400
Hickory	4580	843	6,100
Oak (Red)	3470	706	4,400
Oak (White)	3560	671	4,700
Ash (White)	4060	860	5,300
Cherry	3540	440	4,200

STRUCTURAL WOOD PRODUCTS

Today's designers have a wide range of wood-based materials to choose from. Although solid lumber cut as timbers, dimension lumber, or boards still represent a significant percent of structural wood products, technology and innovation have provided many *composite* or *reconstituted* wood products. Composite or reconstituted wood products are materials produced by combining wood chips or flakes with a binder. The binder is usually a polymer-based glue.

These products are generally a response to changing timber resources and a need for structural products that meet specific operational conditions. New products are engineered to increase stability of properties, therefore improving the wood product's reliability.

STRUCTURAL COMPOSITE LUMBER (SCL)

Structural composite lumber (SCL) is a fairly new product. SCL products are formed by gluing together veneers or strands. Several types of products manufactured in sizes and shapes similar to sawn lumber are available. SCL was developed when high-quality lumber from forest resources started getting hard to find.

A type of SCL is *laminated veneer lumber (LVL).* LVL is made from specially graded veneers. *Veneers* are thin layers of wood (1/10" to 3/16") peeled from a log, then carefully matched and glued together to form sheet stock. A variety of end joints are staggered as the veneers are laminated. LVL is either 2' or 4' wide. The length can be unlimited. However, in practice the length is 80' or less since this is the length of a semitrailer.

Parallel strand lumber (PSL) is another type of SCL. PSL is created by gluing strands of wood or strips of veneer together under high pressure and temperature. The properties of the resulting product are determined by the properties of the woods used in its manufacture.

STRUCTURAL PANEL PRODUCTS

Plywood and particle board have been among the more popular structural panel products for house construction for decades. Today, structural panel products also include waferboard, oriented strand board, and veneer-particle composite panels. These products are available in 4' x 8' sheets in thickness of 1/4" to 1".

Plywood

Plywood is perhaps the best known type of structural panel. It is constructed of veneer, or plies. The veneer is created by rotating a log against a sharp blade, Figure 9-19. After the veneers are graded, they pass through a glue spreader. The veneers are stacked in cross layers with their grain directions at right angles to one another. The arrangement of the plies at right angles provides strength, stability, and minimizes dimensional changes. The stack of veneers is then pressed and heated to cure the glue. Once cured, the panels are trimmed to finished size and prepared for shipping.

Figure 9-19.
Veneer is created by rotating a log across a blade. A thin veneer is shaved from the log.

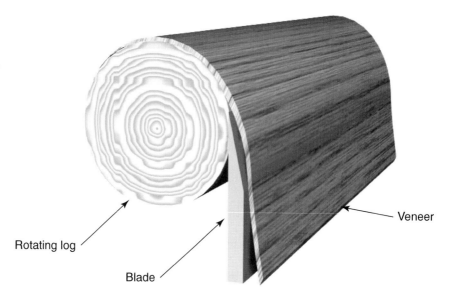

Rotating log

Blade

Veneer

Figure 9-20.
These are two examples of stamps from the American Plywood Association.

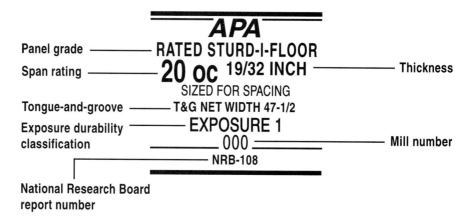

Panel grade — **APA RATED STURD-I-FLOOR**

Span rating — **20 OC** 19/32 INCH — Thickness

SIZED FOR SPACING

Tongue-and-groove — T&G NET WIDTH 47-1/2

Exposure durability classification — EXPOSURE 1

000 — Mill number

NRB-108

National Research Board report number

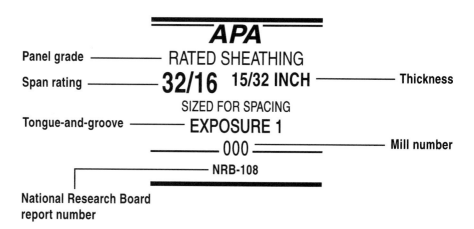

Panel grade — **APA** RATED SHEATHING

Span rating — **32/16** 15/32 INCH — Thickness

SIZED FOR SPACING

Tongue-and-groove — EXPOSURE 1

000 — Mill number

NRB-108

National Research Board report number

Plywood is classified as interior and exterior. *Interior plywood* is made for use inside. It is bonded with adhesives that are partially water soluble. Water soluble adhesives deteriorate or dissolve if exposed to excessive moisture. *Exterior plywood* is made for use outside. It is bonded with adhesives that are 100% waterproof. Phenolic and resorcinol adhesives are commonly used in exterior plywood.

Plywood is manufactured to defined minimum standards. Structural plywood is made to conform to *U.S. Product Standard PS-1.* This standard defines acceptable wood species, veneer grades for individual plies, and composition of the individual plies in a sheet. Plywood sheets are stamped with *The Engineered Wood Association (APA)* registered trademarks, Figure 9-20. These stamps provide important information regarding panel grades, span rating, exposure durability classification, and mill number.

The veneers have a letter grade. *Grade N* is the best and usually a special order. It is primarily used for applications where a natural finish is required. *A, B, C, D* grades are the other veneer grades. Of these, A is the best and D the worst. Grade A means the veneer is smooth and paintable, with no more than 18 neatly made repairs. Grade D means the veneer can have knots and knotholes to 2-1/2″ across. A piece of plywood marked A-D has one face veneer of A quality and one face veneer of D quality. The interior plies of a piece of plywood may be of any grade veneer. However, they are typically of D grade.

Southern pine is a very common softwood used to make plywood. Other species include Douglas fir, western birch, western hemlock, Sitka spruce, commercial white firs, cedar, California redwood, yellow poplar, and sweetgum. The more popular hardwood species used to make plywood are oak, mahogany, birch, cherry, walnut, and maple.

Nonveneer Structural Panels

Waferboard, oriented strand board, and particle board are made from wood particles, strands, and flakes, rather than veneers. Therefore, they are called *nonveneer structural panels.*

These products were developed in response to several factors. First, the availability logs suitable for making plywood veneers has decreased. Second, old growth, large logs are becoming more difficult to find. Third, nonveneer panels are less expensive. Nonveneer panels are also easier to manufacture and may be manufactured closer to the point of use.

Waferboard

Waferboard is made of rectangular flakes of aspen bonded with powdered phenolic resin. The flakes range in size from 1/8″ to 3″ and are randomly placed in the panel. The flakes and resin are heated and pressed to form the sheet. Waferboard has less bending strength and stiffness than plywood.

Waferboard is a structural panel typically used for sheathing and utility applications, Figure 9-21. It is available in thickness of 1/4″ to 3/4″, widths, and lengths similar to plywood. It can be manufactured in sizes up to 8′ x 24′. It often has tongue-and-groove edges.

Figure 9-21.
A sample of waferboard.
Notice the wood flakes
and chips.

Oriented strand board (OSB)

Oriented strand board (OSB) was developed as a structural panel with greater strength and stiffness than waferboard. The strands of wood used in OSB are longer and narrower than those in waferboard. The strands are typically 2″ to 3″ long and 1/4″ to 1/2″ wide. The strands are oriented in the panel for strength. The outer strands typically run parallel to the length of the panel and the inner strands perpendicular to the length. The strands in OSB are bonded with phenolic resin adhesive and hot pressed to form the panels. OSB is made in thickness of 1/4″ to 3/4″.

SUMMARY

Wood has been used by humans for thousands of years for heat, shelter, transportation, furniture, and recreation. It has many advantages over other manufacturing materials and it remains the material of choice for many applications. Wood and manufactured wood products are popular materials for residential construction.

Wood is aesthetically pleasing and warm. As a resource, it is renewable, easily shaped, strong, and noncorrosive. Although older, mature forests are diminishing, conservation and reforestation are helping ensure a supply of quality wood.

KEY TERMS

A Select
Ablation
American Plywood
    Association (APA)
Anisotropic
Annual Rings
B and Better
B Select
Bending Strength
Boards
C Select
Cambium Layer
Cellulose
Central and Southern
    Forest Region
Common Grades
Composite
Coniferous
Conservation
Construction
D Select
Debarking
Deciduous
Density
Dimension
Early Wood
Equilibrium Moisture Content
Exterior Plywood
Extractives
Factory and Shop
Fiber Saturation Point
Firsts and Seconds (FAS)
Grade A
Grade B
Grade C
Grade D
Grade N
Green
Heartwood
Hemicellulose
Hygroscopic
Inner Bark
Interior Plywood

Laminated Veneer
    Lumber (LVL)
Light Framing Lumber
Lignin
Log
Macrostructure
Medullary Rays
Microfibril
Microstructural Level
Modulus of Rupture
No.1
Nonveneer Structural Panels
Northern Forest Region
Oriented Strand Board (OSB)
Outer Bark
Oven Dried Method
Parallel Strand Lumber (PSL)
Phloem
Pith
Plain-Sawing
Plywood
Quartersawing
Reconstituted
Redwood Region
Reforestation
Sapwood
Sawmill
Seasoning
Select Grades
Southern Pine Region
Specific Gravity
Springwood
Standard
Structural Composite
    Lumber (SCL)
Summerwood
Timber
Timbers
Tree Farm Certification
U.S. Product Standard PS-1
Utility
Western Wood Region
Yard Lumber

QUESTIONS FOR REVIEW AND DISCUSSION

1. List five advantages of wood as a manufacturing or construction material.

2. What four parts make up wood? What portion of the volume of wood does each represent?

3. Define the following tree parts.

Pith	Heartwood
Cambium layer	Outer bark
Medullary rays	Sapwood
Inner bark	

4. In a cross section of a tree trunk, how can you identify springwood and summerwood?

5. List advantages and disadvantages of quartersawn lumber.

6. List the advantages and disadvantages of plain-sawn lumber.

7. What is fiber saturation point?

8. What is equilibrium moisture content?

9. How does moisture content affect properties of wood?

10. What is the moisture content of seasoned lumber?

11. What subclassifications are used for softwood lumber?

12. What is the difference between density and specific gravity?

13. Define bending strength.

14. List and describe four reconstituted wood products.

FURTHER READINGS

1. Asheland, Donald R. The Science and Engineering of Materials (3rd Edition). PWS: Boston (1994).

2. Wood Handbook: Wood as an Engineering Material. US Department of Agriculture, Forest Services.

INTERNET RESOURCES

www.fpl.fs.fed.us
   Forest Products Laboratory

www.apawood.org
   The Engineered Wood Association

www.southernpine.com
   Southern Pine Council

www.natlhardwood.org
   National Hardwood Lumber Association

www.wwpa.org
   Western Woods Product Association

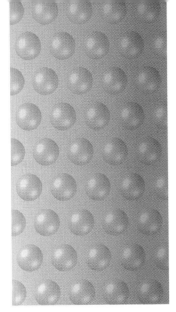

10

Ceramics and Glass

KEY CONCEPTS

Upon completion of this chapter, you should understand:
- ➢ Historical development of ceramics
- ➢ Structure of ceramics and glass.
- ➢ Composition of ceramics.
- ➢ Classification of ceramic products.
- ➢ Uses of ceramic compound by classification.
- ➢ Physical properties of ceramics.
- ➢ Forming processes used with ceramics.
- ➢ Post forming processes used with ceramics.

It is difficult to pinpoint an exact date when ceramics were first discovered and used. However, evidence suggests that ceramics were developed as a result of the discovery and use of fire. Perhaps as early as the Neolithic Age, about 10,000 years ago, our ancestors discovered that ground beneath a cooking fire at times became hard. Eventually, they applied this observation in a practical way by forming and heating clay to make containers for storing food and water.

When most people hear the word "ceramics" today, they think of a potter's wheel and handmade vases and bowls. However, ceramics have played a very important role in industrialization. Their unique properties and availability make them the materials of choice for a wide variety of applications and products. Ceramics can be found in many products. Certain ceramics are used in foundations of houses and commercial buildings, where they exhibit tremendous strength. Other types of ceramics are used on the space shuttle fuselage to protect the shuttle and its passengers from high re-entry temperatures, Figure 10-1.

This chapter explores the composition, properties, and manufacturing processes of modern industrial ceramics. The uses for traditional and advanced ceramics are growing substantially. In fact, some have predicted that ceramics will be to the twenty-first century what plastics were to the second half of the twentieth century.

Figure 10-1.
The ceramic heat shield tiles on the underside of the space shuttle orbiter. (NASA)

INTRODUCTION

Ceramics are defined as hard, brittle compounds of metallic and nonmetallic elements that have high melting temperatures and are chemically inert. The term ceramics is derived from the Greek word "keramos." The literal translation of this word means burnt earth. Ceramics are composed of the most abundant elements on earth, such as silica, alumina, and magnesia. Heat is used to finish the ceramic. This is called *firing*. Ceramics are often broadly classified as crystalline or noncrystalline.

Crystalline ceramics have orderly arrangements of their ions or atoms which form lattice structures of repeating patterns, Figure 10-2. Most traditional ceramics, such as oxides, nitrides, and carbides, have crystalline structures. Most ceramics are crystalline solids consisting of either single crystals or multiple crystals which are referred to as polycrystalline solids. A single crystal is a solid in which the periodic and repeated arrangements of atoms is perfect and extends throughout the entirety of the ceramic body without interruption. A polycrystalline solid is a compound of a collection of many single crystals separated from one another by grain boundaries, Figure 10-3.

Noncrystalline ceramics are vitreous and their atomic or ionic structure is not ordered. Vitreous means low porosity. Glass is the classic example of a noncrystalline vitreous ceramic.

Figure 10-2.
Crystalline ceramics have
repeating unit cell structures.

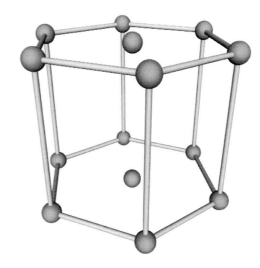

Figure 10-3.
The tetrahedral structure
of ceramic silicates.

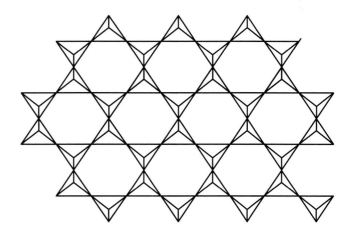

COMPOSITION

Ceramics are generally composed of more than one element. They are often composed of metallic and nonmetallic elements. Ceramics containing more than one earth element are referred to as *polyphase ceramics.* Silicon carbide (SiC) and aluminum oxide (AL_2O_3) are examples of polyphase ceramics. Ceramics may also be composed of a single element. These are called *monophase ceramics.*

Traditional ceramics are made from abundant, naturally occurring earth elements that form what is commonly referred to as clay. The principal elements found in these clays are oxygen, silicon, and alumina. These elements form compounds known as *alumina silicates.* Oxygen found in the metal oxides of ceramic compounds constitutes 46.6% of the earth's crust, silicon 27.7%, and alumina 8.1%. Other high volume elements that are found in ceramics are iron and magnesium. Additionally, the nitrides, borides, carbides, and silicides of all metals are, by definition, ceramics.

Often, quartz and feldspar are added to clay to improve its properties. Next to clay, feldspar is the most important ingredient in many ceramic products. Feldspar is the universal *flux* used in all types of ceramic bodies. A flux cleans the clay of oxides and other impurities. There are many varieties of feldspars. The most important commercial feldspars are potash feldspars, othoclose, soda feldspar, albite and anorthite. Aplite is another flux used for ceramics and has a lower volume of alumina and more silica.

The composition of clay varies according to the application. For example, clay used to manufacture bricks is 10% magnesia, 15% lime, 8% iron oxide, 10% to 25% alumina, and 35% to 65% silica. Most clays contain quartz, sand, and some mica. The following is a list of the more commonly used clays in traditional ceramics.

- **Kaolin.** This is the purest form of clay, often referred to as China clay.
- **Ball Clay.** A sedimentary clay of excellent plasticity, used extensively in manufacturing whiteware.
- **Fireclay.** High melting temperature clays commonly used for lining furnaces, flues, and firebrick. These clays contain alumina, silica, iron oxide, and lime.
- **Slip Clay.** A low purity clay used as a glaze on ceramic products.
- **Flint Clay.** A hard, low plasticity clay used for mixing in ceramics to reduce drying and firing shrinkage.

CLASSIFYING CERAMICS

Although both natural and synthetic ceramics are available, ceramics are usually classified by use or product type, Figure 10-4. The following list is a commonly used method of classifying ceramics by use.

- Structural clay
- Refractories
- Porcelain enamel
- Whiteware
- Glass
- Advanced (engineering ceramics)

STRUCTURAL CLAY

Structural clay represents approximately 2% of the total annual sales in the ceramics industry. Structural clay products include brick, drain tile, flue linings, and floor tile, Figure 10-5. These products are manufactured from the most abundant, lowest-cost clays. Clays used for structural ceramics are usually extracted locally and can be fired at relatively low temperatures. Structural clay products are ceramic bodies used where strength is required to support a force or load.

Structural clays were among the first ceramics used by humans. Bricks were manufactured and used as early as 3000 BC by the Samerians in Mesopotamia. They used the bricks to build temples and fortifications, which gave them supremacy over their neighbors.

Figure 10-4.
This pie chart shows the use of ceramics by sales percentage.

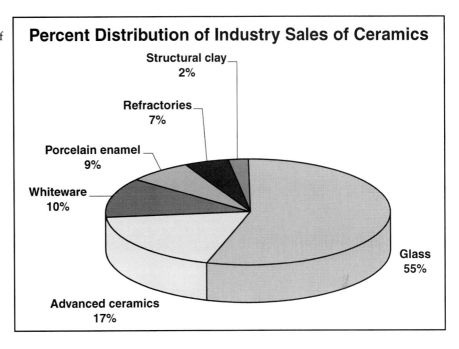

Percent Distribution of Industry Sales of Ceramics

Structural clay
2%

Refractories
7%

Porcelain enamel
9%

Whiteware
10%

Glass
55%

Advanced ceramics
17%

Figure 10-5.
These drain tiles are made of ceramic.

Specifications of the American Society of Testing and Materials (ASTM) covers four types of concrete masonry units. These are listed below.

- Concrete building brick.
- Hollow load-bearing-concrete masonry units.
- Hollow no-load-bearing concrete masonry units.
- Solid load-bearing-concrete masonry units.

REFRACTORIES

Refractory ceramics are high melting point oxides frequently referred to as crystalline ceramics. Most refractory ceramics are alumina and silica mixed in various proportions, dry pressed to shape, and fired to develop strength. Refractory ceramics represent about 7% of the total sales of ceramic products.

Refractory ceramics are used in many applications. They are used in crucibles used for melting metals and for brick linings of large steel making furnaces. See Figure 10-6. These brick linings are also used for smaller furnaces and kilns used to heat treat metals and fire ceramic whiteware.

PORCELAIN ENAMEL

Porcelain enamels are glassy coatings generally applied to kitchen and bathroom appliances and fixtures, Figure 10-7. They are also applied to metal architectural panels and signs. Enamels are fused onto metal substrates to provide protection against corrosion and a pleasing decorative appearance.

Porcelain enamels represent approximately 9% of annual sales in ceramics. The largest consumers of porcelain enamels are the major appliance companies. Approximately 82% of all porcelain enamels are used by appliance manufacturers.

The glass used in enameling is specially prepared to meet various physical requirements. These requirements may include color, opacity, or processing requirements. These specially compounded glasses are ground into fine powder called *frit*. The frit is uniformly applied to the metal product. The product is then placed into a low temperature furnace to permanently fire the enamel to the metal.

Figure 10-6.
This ceramic crucible is used in metal casting.

WHITEWARE

Ceramic whiteware, often referred to simply as *whiteware,* is used to manufacture dinnerware, ceramic bathroom tile, fine china, and earthenware, Figure 10-8. These products have essentially the same compounds as the structural clays. However, the ratio of elements is different. In most cases, the ratio of feldspar to clay is higher.

Figure 10-7.
This pedestal sink has an enamel coating.
(Kull Lumber, Mattoon, IL)

Figure 10-8.
This dining set is whiteware.

Ceramic whitewares are usually formed by slip casting or dry pressing followed by a firing process called vitrification. *Vitrification* is a process which uses high temperature to cause partial melting and produces a dense, hard structure.

Vitrification is followed by a process known as glazing. *Glazing* is applying a glassy coating to seal the surface of the whiteware product. The glaze prevents absorption of liquids and improves its appearance. Pigment may be added to the glaze to change the color of the product.

GLASS

Glass is a noncrystalline ceramic made from a mixture of silica and basic oxides, such as boron. The noncrystalline nature of glass causes it to behave like a very thick liquid. Glass is sometimes called a semisolid.

The basic ingredient in glass is silica. However, the composition of glass is varied by adding other minerals to produce a wide range of properties. There are perhaps as many as 10,000 different compositions of glass.

Glass represents approximately 55% of the annual sales in the ceramics industry. Glass is used to make containers, light bulbs, TV tubes, laboratory glassware, plate glass for windows, and glass fibers for insulation and reinforced plastics. Flat (plate) glass represents the largest segment of the glass industry, about 32% of sales.

The four most popular commercial glasses are lead, soda-lime, borosilicate, and high silicate, Figure 10-9. Soda-lime is the glass most people are familiar with. It is used for everyday products including bottles, light bulbs, and windows.

ADVANCED (ENGINEERING CERAMICS)

Advanced (engineered ceramics) is a classification used to define a group of ceramics tailored or engineered to meet specific operational conditions. These compounds are typically used for high technology applications.

Figure 10-9.
This table shows the properties of various types of commercial glass.

COMMERCIAL GLASS COMPOSITION (WT %)								
Type	Composition							
	SiO_2	B_2O_3	AL_2O_3	Na_2O	K_2O	MgO	CaO	PbO
Soda Lime	70		1	15		4	10	
Soda Silica	72		1	20		3	4	
Lead Silica	63		1	8	6		1	21
Fused Silica	99.8			5		0.1	0.1	
Borosilicate	76	13	4				1	

Advanced, or engineered, ceramics are designed to meet the growing demand for special applications in electronics, transportation, nuclear power, and communication. Optical fibers, capacitors, electrical porcelain, integrated circuits, and electrical insulators are among the many products manufactured from advanced ceramics. Advanced ceramics represent about 17% of the annual sales of ceramics. It is one of the fastest growing segments of the ceramics industry.

Advanced ceramics are made of extremely high purity oxides that exhibit excellent chemical and electrical properties. Examples of advanced ceramics include silicon carbide (SiC), silicon nitride (Si_3N_4), zirconia (ZrO_2), alumina (AL_2O_3), tungsten carbide (WC), boron carbide (B_4C), and boron nitride (BN). Figure 10-10 illustrates the more common engineered ceramics and their applications.

STRUCTURE

As compounds, ceramics consist of various types of atoms of varying sizes. The atoms of the metallic and nonmetallic elements are attached to one another by ionic or covalent bonds. The attraction between positively charged metal atoms and negatively charged nonmetallic atoms form an extremely tight bond. This bond produces unique properties of ceramics.

Many ceramics are made of silicates. Silicates are compounds of silicon and oxygen. The silica structure has an integral arrangement consisting of tetrahedral units, Figure 10-11. In this structure, four oxygen atoms are covalently bonded to each of the smaller silicon atoms. Four valence electrons are shared. The silicate (SiO_4) tetrahedral structure is the building block of silicon-based covalent ceramics. These ceramics which include brick, cement, and whiteware. Concrete and cement blocks are essentially silicates ceramics as well.

Figure 10-10.
This table shows the applications and properties of advanced ceramics.

Ceramic	MOR (MPa)	Comp. Strength (MPa)	Fracture Toughness (MPa) $m^{-1/2}$	Therm Exp. Coefficient m/m/k (0-1000°C)	Applications
PROPERTIES AND APPLICATIONS OF ADVANCED CERAMICS					
Alumina (AL_2O_3)	380	2750	1.75	9.0×10^{-6}	Electrical insulators, abrasives
Zirconia (ZrO_2)	610	1850	9.5	10.6×10^{-6}	Refractory ceramic
Silicon Carbide (SiC)	483	2000	3	4.3×10^{-6}	Abrasives
Tungsten Carbide (WC)	1378	4823	6.0–20.0		Cutting tools
Boron Nitride (BN)	48–100	310		4.4×10^{-6}	Insulators lubricant
Boron Carbide (B_4C)	340	2760		5.5×10^{-6}	Abrasives

Figure 10-11.
This is the polycrystalline
structure of a ceramic body.

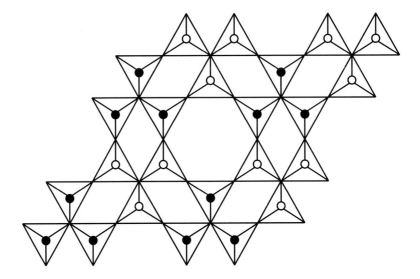

Many different structures can be formed by bonding silicon and oxygen atoms in a variety of configurations. These configurations include:
- silica sheets (clays, talcs, mica)
- chain silicates (asbestos)
- island silicates (garnets olivines)
- aluminosilicates (clays)

PROPERTIES

The properties of ceramics are a function of composition and structure. As a class, ceramics are hard, brittle materials that do not readily conduct heat or electricity. Ceramics have high melting temperatures, high modulus of elasticity, high hardness, high compressive strength, low electrical and thermal conductivities, and low impact strength. For example, the high compressive strength of concrete makes it an excellent choice for columns and footings in buildings. In addition, the thermal properties of ceramics make them excellent insulators.

MECHANICAL PROPERTIES

Mechanical properties are the characteristics of the ceramic body displayed when a force is applied to it. The mechanical properties of ceramics that are important include tensile strength, compressive strength, flexural strength, stiffness, facture toughness, and hardness.

Tensile and Compressive Strength

Ceramics, especially concrete, have been traditionally used in applications where resistance to compressive forces are required. Generally, the compressive strength of ceramics is many times greater than their tensile strength. This is due to the microstructure of ceramic materials. Ceramics are composed of various-size particles. When fused together, voids and small cracks are produced. In compression, these voids have a tendency to fuse together and resist failing. In tension, however, these voids are weaknesses where failure occurs.

Because of their brittle nature, ceramics do not plastically deform. Therefore, they do not have a yield strength. They do not exhibit permanent elongation or reduction due to plastic deformation.

Theoretically, ceramics have high tensile strengths. It is possible to achieve tensile strengths of 700,000 psi to 1,400,000 psi (4827 MPa to 9655 MPa) from newly drawn glass fibers. However, in practice, these strengths cannot be achieved. The smallest flaw in the glass fiber dramatically decreases the tensile strength.

Tensile testing ceramic bodies is both difficult and expensive. Machining ceramic specimens to a standard size and shape is time consuming and requires costly tooling. Therefore, most information related to tension is achieved through flexural tests.

Flexural Strength

In flexural tests, a beam of standard size is placed between two points and a force is applied perpendicular to the beam axis, Figure 10-12. In this test the lower surface of the beam experiences tension forces.

The flexural strength of ceramics is calculated by using the formula for modulus of rupture.

$$MOR = \frac{3FL}{2BH^2}$$

where F is the applied force, L is the length of the beam between supports, B is the width of the test specimen, and H is the thickness of the specimen. Figure 10-13 shows flexural strengths of selected ceramics.

Figure 10-12.
The setup for a flexural test of a ceramic beam.

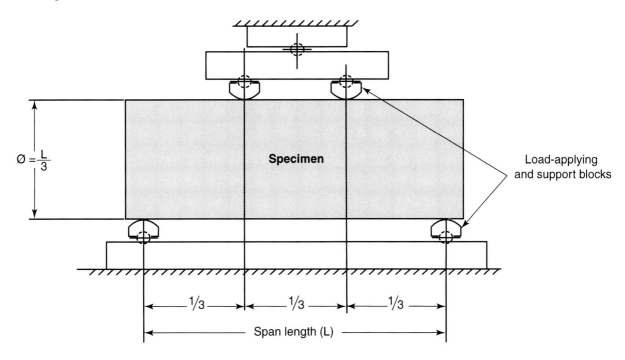

$\varnothing = \frac{L}{3}$

Specimen

Load-applying and support blocks

1/3   1/3   1/3

Span length (L)

Stiffness

Ceramics have the highest modulus of elasticity of all industrial materials. In other words, they are the stiffest of all engineering materials. Advanced ceramics can achieve a modulus of elasticity three times greater than steel. The modulus of elasticity can be as high as 65 million psi, compared to 29 million psi for steel. The formula for modulus of elasticity is:

$$\text{Modulus of Elasticity} = \frac{\Delta\text{Stress } (\sigma)}{\Delta\text{Strain } (E)}$$

Physical conditions affect the modulus of elasticity. For example, temperature can have an effect on the stiffness of some ceramics.

Fracture Toughness

Perhaps the greatest weakness of traditional ceramics is its brittle nature. Traditional ceramics have the lowest fracture toughness value of industrial materials. They typically do not exceed a fracture strength of 50,750 psi (350 MPa). Although progress is being made by adding fiber reinforcement, ceramics are still not the material of choice in applications where mechanical vibration or impact forces are present.

Hardness

Ceramics are the hardest of all industrial materials. The hardest natural ceramic is diamond. Atoms of ceramic materials have strong covalent and ionic bonds, which are difficult to separate. The structure and bonds form an extremely hard material that, for example, resists wear. As such, ceramics make some of the best abrasive and cutting materials. Aluminum oxide (AL_2O_3) and silicon carbide (SiC) are perhaps the most popular abrasives in use today. Most abrasive paper, or "sandpaper," is made with aluminum oxide. Tungsten carbide (WC) is one of the most popular materials for blades and cutting tools. Figure 10-14 illustrates the hardness of commonly used ceramics compared to nonceramic materials.

Figure 10-13.
This table shows the flexural strength of various ceramics.

FLEXURAL STRENGTH OF MATERIALS	
Material	**MOR (Mpa)**
Alumina (AL_2O_3)	380
Zirconia (ZrO_2)	610
Silicon Carbide (SiC)	483
Silicon Nitride (SiN)	241
Boron Carbide (B_4C)	340
Tungsten Carbide (WC)	1378

HARDNESS RATINGS OF MATERIALS	
Material Type	**Vickers Hardness (psi)**
Aluminum Oxide (AL_2O_3)	1900
Silicon Carbide (SiC)	2550
Zirconia (ZrO_2)	1300
Amorphous Glass (SILICA)	600
Silicon Nitride (Si_3N_4)	1750
Tungsten Carbide (WC)	1600
1020 Carbon Steel	120
4340 Molybdenum Steel	220
T11302 Tool Steel	630
A48 Class 20 Gray Cast Iron	175
SA3000 Stainless Steel	260
A96061 (T6061) Aluminum	95
A91100 (1100-H14) Aluminum	26
Be Copper	410
ETP Copper	60

PHYSICAL PROPERTIES

Physical properties are defined as characteristics of materials resulting from interaction with various forms of energy. In the case of ceramics, the energy is usually in the form of heat or electricity. As a family, ceramics are poor conductors of heat and electricity.

Thermal Conductivity of Ceramics

Thermal conductivity of a material refers to its ability to transfer heat from a body of higher temperature to one of lower temperature. The ability of a material to conduct heat is a product of its ability to transfer electrons from one atom in the material to another. The more a material resists this electron transfer, the lower the thermal conductivity. The electrons in ceramics resist flowing from one location to another because of the tight ionic and covalent bonds. Therefore, ceramics typically have low thermal conductivity. Thermal conductivity (K) of some popular industrial materials including ceramics are shown in Figure 10-15.

Figure 10-15.
This table shows the thermal
conductivity of various
materials.

THERMAL CONDUCTIVITY OF MATERIALS	
Material Type	Thermal Conductivity (BTU ft/hr ft^2°F)
Aluminum Oxide (AL_2O_3)	20
Silicon Nitride (Si_3N_4)	19
Silicon Carbide (SiC)	35
Zirconia (ZrO_2)	3.5
Amorphous Glass	0.8
Tungsten Carbide (WC)	50
Polyethelene (PE)	2.28
Polyvinyl Chloride (PVC)	1.04
Polypropylene	1.28
Nylon	1.2
G10100 (1040) Carbon Steel	27
S43000 (430) Stainless Steel	15.1
ETP Copper	226
DHP Copper	196
Tungsten	96.6

Thermal Expansion

Thermal expansion is the degree to which a material increases in volume with an increase in temperature. As the temperature of a material increases, its atoms vibrate. This, in turn, increases the distance between lattice sites. Thus, the material expands with increased heat and contracts with decreased heat. Ceramics as a family of materials generally have the lowest thermal expansion of all materials, Figure 10-16.

The measure of thermal expansion is the linear coefficient of thermal expansion, alpha (α). This is determined from the following formula.

$$\alpha = \frac{\Delta l}{l_o \Delta T}$$
$$\Delta l = l_f - l_o$$
$$\Delta T = T_f - T_o$$

Where l_f is the final length, l_o is the original length, T_f is the final temperature, and T_o is the original temperature.

Figure 10-16.
This table shows the linear coefficient of thermal expansions for various materials.

LINEAR COEFFICIENTS OF THERMAL EXPANSION FOR VARIOUS MATERIALS		
Material	**10^{-5} IN./IN./°F**	**10^{-5} m/m/°K**
Aluminum Oxide (AL_2O_3)	4.1	7.4
Silicon Carbide (SiC)	2.5	4.5
Silicon Nitride (Si_3N_4)	1.7	3.1
Zirconia (ZrO_2)	4.7	8.5
Amorphous Glass	0.55	1
Tungsten Carbide (WC)	4.1	7.4
G10100 (1040) Carbon Steel	8.4	15.2
T41907 (S7) Tool Steel	6.9	12.59
A41 Class 20 Gray Iron	6	10.8
A96061 (6061-T6) Aluminum	13.1	23.6
ETP Copper (C11000)	9.8	17.7

Electrical Conductivity

Electrical conductivity can be defined as the ability of a material to transport electrons. The electrical resistivity of a material is its ability to prevent electron flow. As a class of materials, ceramics do not readily conduct electricity and, therefore, have high resistivity. This makes ceramics excellent electrical insulators. Figure 10-17 shows the electrical resistivity of traditional and engineering ceramics compared to other industrial materials. The lower the resistivity the higher the electrical conductivity.

Nonoxide ceramics (SiC) contain transition metal ions. This makes these ceramics extremely important as semiconductors.

FORMING CERAMICS

Post processing of ceramics is both difficult and expensive. Because ceramics are so hard and brittle after they are fired, expensive tooling and time consuming processes are required to machine them. Therefore, most forming processes are designed to produce a finished product or one that requires little post processing. When post processing is required, grinding is generally the method used.

Figure 10-17.
This table shows electrical
resistivity for various
materials.

ELECTRICAL CONDUCTIVITY OF MATERIALS	
Material	Electrical Resistivity 10^{-6} Ohms/cm
Aluminum Oxide (AL_2O_3)	10^{12}
Silicon Carbide (SiC)	0.1
Silicon Nitride (Si_3N_4)	10^{10}
Zirconia (ZrO_2)	10^8
Amorphous Glass (SILICA)	10^7
Cemented Carbide (Wc)	6×10^{-8}
G101000 (1010) Carbon Steel	20×10^{-6}
A48 Glass 20 Gray Iron	100×10^{-6}
Pure Nickel	11×10^{-6}
Epoxy	6.1×10^{15}
Polystyrene	10^{16}
Polyethylene	10^{18}

All forming processes begin with careful preparation of powders. In traditional ceramics, silica, clay, fluxes, and refractory materials in powder forms are carefully blended. In addition to this, for advanced ceramics the purity and particle size of the various powders is extremely critical. The transformation of the carefully processed powders into desired shapes is usually accomplished by one of the following green forming processes. Green forming means the clay contains moisture making it plastic and, therefore, easily deformed.

  ◆ Casting
  ◆ Extrusion
  ◆ Jiggering
  ◆ Die Pressing
  ◆ Isostatic Pressing
  ◆ Tape Casting
  ◆ Injection Molding

CASTING

Most ceramic products formed by casting are produced by one of two methods. These methods are slip casting and solid casting.

Slip casting is used to form hollow shapes, such as toilets and sinks. *Slip* is a low-viscosity ceramic slurry. During slip casting, slip is poured into a plaster of paris mold. The mold is in the shape of the desired ceramic product. Capillary action of the porous plaster mold draws water from the slip. This causes the slip to form a stiff, or solid, ceramic layer on the mold surface. The longer the slip is allowed to remain in the mold, the thicker the walls of the product become. When the desired thickness is obtained, the remaining slip is poured from the mold leaving the desired product, Figure 10-18.

In *solid casting,* the process is similar to slip casting. However, the mold includes riser columns to feed the slip as capillary action occurs. The risers feed the casting as it loses moisture and begins to shrink, Figure 10-19.

EXTRUSION

During *extrusion,* an uncured ceramic body is forced through a die. The die has the desired cross section of the finished product. The extruded ceramic is fed onto a conveyor belt at the end of the die head where it is cut to length, Figure 10-20. This process is used to form structural ceramics. Generally, any ceramic product with a continuous cross section can be formed in this manner. Bricks, pipes, ceramic tubing, and some electrical ceramic components are formed by extrusion.

Figure 10-18.
Slip casting of ceramics can be used to create ceramic jars, vases, and containers.
A—Slip is poured into the mold. B—Water is absorbed by the mold. C—Excess slip is removed. D—The finished, hollow part (shown partially transparent for illustration).

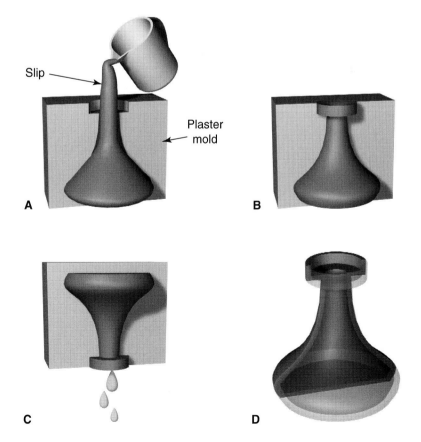

Figure 10-19.
Solid casting of ceramics can
be used to create solid ceramic
products. A—Slip is poured
into the mold. B—The ceramic
solidifies. C—The finished
part.

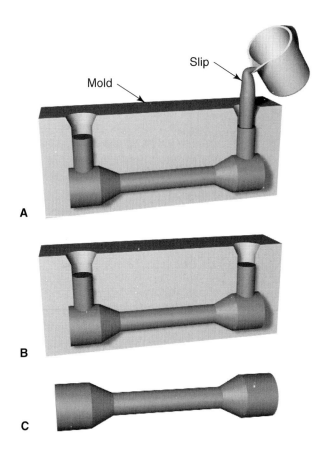

Figure 10-20.
Ceramics can be extruded
using a screw extruder.

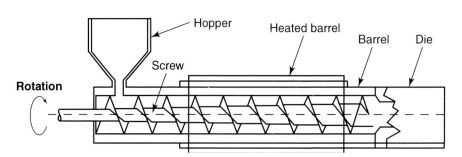

JIGGERING

Jiggering involves placing an uncured ceramic body (clay) between matching halves of a mold, Figure 10-21. The clay and one-half of the mold rotate as the other half of the mold is forced against them. The force and rotation causes the clay to conform to the contour of the mold. Dinner plates are perhaps the best example of a product made using this process.

Figure 10-21.
Jiggering involves placing an uncured ceramic body (clay) between matching halves of a mold. One-half of the mold applies pressure while the other half rotates to produce the finished part.

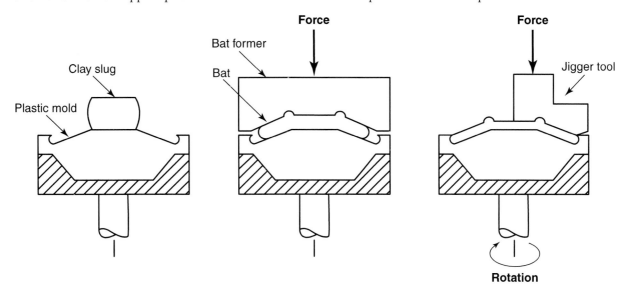

Figure 10-22.
Die pressing of ceramics involves heat and pressure.

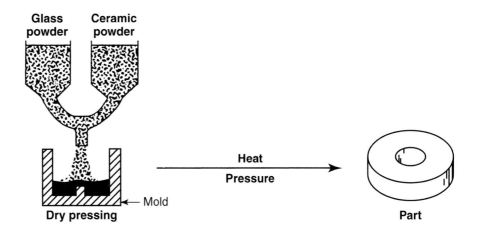

DIE PRESSING

A type of powder pressing is used for a variety of products, including ceramic floor tile, ceramic wall tile, and many electronic components. *Die pressing* uses steel or carbide dies and hydraulic force, Figure 10-22. The uncured clay body is placed in a die cavity in the shape of the desired product. Hydraulic force is applied to compact the clay and conform it to the die cavity.

Powder pressing uses fine ceramic powders forced together under tremendous loads and then heated. This causes the powder to stick together.

ISOSTATIC PRESSING

    Isostatic pressing is a forming process used for more complex shapes, Figure 10-23. Isostatic pressing uses a pressure vessel to apply fluid pressure to the outside of the mold. The powdered clay is blended with additives, such as organic binders and lubricants, to improve the ability to flow into complex die cavities.

TAPE CASTING

    Tape casting involves drawing a powdered slurry under a blade to form a thin layer on a moving plastic belt, Figure 10-24. Evaporation removes water from the slurry and a ceramic "tape" is formed. This is the primary process used to form ceramic substrates for integrated electronic circuits.

Figure 10-23.
Isostatic pressing uses a pressure vessel to apply fluid pressure to the outside of the mold.

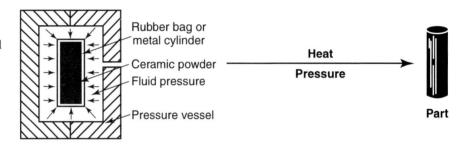

Figure 10-24.
Tape casting involves drawing a powdered slurry under a blade, or gate, to form a thin layer on a moving plastic belt.

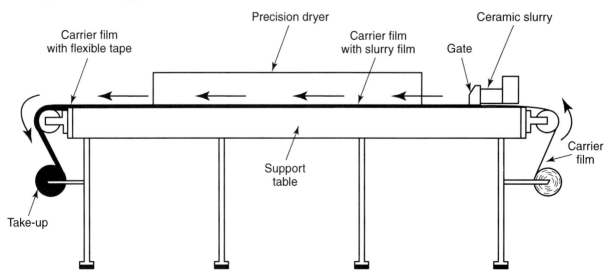

INJECTION MOLDING

Injection molding of ceramics is similar to injection molding of plastics. A blend of ceramic powders is forced by a cylindrical plunger or screw through a barrel and into a steel mold, Figure 10-25. The mold has a cavity the shape of the desired product. Injection molding is used to form complex solid shapes.

POST FORMING PROCESSES

The forming processes described in the previous sections produce *green ceramic* shapes. These shapes have little stiffness. To complete the processing, drying is usually required. Drying processes vary depending upon the forming process used. The purpose of *drying* is to remove moisture from the green ceramic body before densification takes place. *Densification* involves reducing the porosity of the ceramic body, thus forming a dense part. Densification is a three step process involving sintering, hot pressing, and hot isostatic pressing (HIP).

Sintering involves applying a high temperature that causes a viscous flow of powders to reduce porosity and cause densification. Sintering is usually referred to as *firing* or vitrification.

Hot pressing is used to densify ceramics that cannot be sintered. Hot pressing involves the use of a die, pressure, and heat to reduce porosity of the ceramic body, thus increasing its density. The die is usually made from graphite.

Hot isostatic pressing (HIP) is similar to isostatic pressing used in forming complex ceramic bodies. However, in HIP, the pressing is done at higher temperatures. Also, gas is used as the pressurizing fluid instead of liquid.

Figure 10-25.
With injection molding of ceramics, a blend of ceramic powders is forced by a cylindrical plunger or screw through a barrel and into a steel mold.

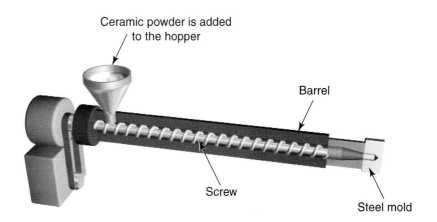

Ceramic powder is added to the hopper

Barrel

Screw

Steel mold

SUMMARY

Ceramics are one of the oldest materials used by humans to extend their potential. Early humans discovered that certain parts of the ground harden and retain shape when heated. This discovery led to the forming of ceramic vessels for storage of water and foodstuff, and later to the manufacture of bricks and other structural ceramics for fortification and shelter. Ceramics are composed of common earth elements, such as oxygen, silicon, alumina. These elements constitute more than 80% of the earth's crust.

While art is a common application for ceramics, they are extremely important as industrial materials. New applications are being developed rapidly. Ceramics have become critical materials for the electronics industry and play a vital role in the space program.

Ceramics are hard, brittle, crystalline or noncrystalline materials that do not readily conduct heat or electricity. Although they fracture easily, they have high compressive strength. This makes them excellent materials for foundations. Ceramics are the hardest of all materials, which gives them high wear resistance. High wear resistance makes ceramics ideal as abrasives and cutting tools.

IMPORTANT TERMS

Advanced (Engineered) Ceramics	Injection Molding
Alumina Silicates	Isostatic Pressing
Ceramics	Jiggering
Crystalline	Monophase Ceramics
Densification	Noncrystalline
Die Pressing	Polyphase Ceramics
Drying	Porcelain Enamels
Extrusion	Refractory Ceramics
Firing	Sintering
Flux	Slip
Frit	Slip Casting
Glass	Solid Casting
Glazing	Tape Casting
Green Ceramic	Thermal Expansion
Hot Isostatic Pressing (HIP)	Vitrification
Hot Pressing	Whiteware

QUESTIONS FOR REVIEW AND DISCUSSION

1. From what Greek word is the term ceramics derived? What is its literal meaning?
2. What is the definition of ceramic?
3. What is the difference between a crystalline ceramic and a noncrystalline ceramic?
4. What three abundant earth elements are found in most ceramics?
5. What are polyphase ceramics?
6. What is a monophase ceramic?
7. What function does feldspar provide in the manufacture of ceramic materials?
8. List and describe the five commonly used clays for ceramics.
9. List the six classifications of ceramics and provide examples of products for each.
10. Describe the basic arrangements of atoms in the silicate lattice structure.
11. What four structures are typical of the bonding of silicon and oxygen atoms?
12. How does a ceramic typically perform under tension?
13. Why is tensile testing of ceramics difficult?
14. How is tensile testing data for ceramics obtained in practice?
15. A ceramic beam is tested to determine its modulus of rupture (MOR). The applied force is 30,000 psi, the span length of the beam is 28″, its width is 2″, and its thickness is 2″. Compute the MOR.
16. What is used as a measure of stiffness in ceramics?
17. How is the fracture toughness of ceramics improved?
18. What two ceramics are the most popular synthetics for abrasive materials?
19. Explain how thermal expansion occurs.
20. A brick 8″ long and 72°F is placed in a 500°F oven. The brick is left in the oven until it reaches 500°F. At 500°F , the length of the brick is 8.15″. What is the linear coefficient of thermal expansion for the brick?
21. Why is post processing of ceramics kept to a minimum?
22. Describe slip casting of ceramics.
23. Describe solid casting of ceramics.
24. What process is used to forming whiteware plates and saucers?
25. What is done to reduce how porous the ceramic body is after the forming process?

FURTHER READINGS

1. Barasum, Michael. <u>Fundamentals of Ceramics</u>. New York: McGraw-Hill Inc. (1997).

2. Brady, George S. & Clauser, Henry R. <u>Materials Handbook, 13th edition</u>., New York: McGraw-Hill, Inc. (1991).

3. Jones, J. T. and Berard, M.F. <u>Industrial Processing and Testing (2nd edition)</u>. Ames, IA: Iowa State University Press (1993).

4. Reed, James S. <u>Principles of Ceramics Processing (2nd edition)</u>. New York: John Wiley and Sons, Inc. (1995).

INTERNET RESOURCES

www.ceramicindustry.com
  Ceramic Industry Magazine

www.acers.org
  The American Ceramic Society

www.ceramics.com/list.html
  Ceramics and Industrial Minerals

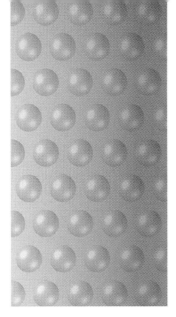

Chapter 11

Cement and Concrete

KEY CONCEPTS

Upon completion of this chapter, you should understand:
- ➢ History of cement and concrete.
- ➢ Difference between cement and concrete.
- ➢ Characteristics and properties of cement and concrete.
- ➢ Chemical process which causes cement and concrete to cure.
- ➢ Manufacturing process used to make Portland cement.
- ➢ Types of Portland cement and their uses.
- ➢ Composition of cement and concrete.
- ➢ Classifications of concrete.

Although the terms "cement" and "concrete" are used synonymously by the general public, they refer to quite different materials. In fact, cement is an ingredient in concrete and other materials. Concrete is an excellent example of a composite material that has been in use for many years. These two materials are defined as follows.

- ◆ *Cement:* A finely pulverized material consisting principally of compounds of lime, silica, alumina, and iron
- ◆ *Concrete:* A monolithic, ceramic product of aggregates bonded with cement, such as Portland or asphalt

In this chapter, the unique characteristics and qualities of both concrete and cement are discussed. These materials represent a large portion of the material used annually in industry.

CEMENT

Early civilizations needed a way to fasten rocks together to construct tombs, fortifications, and temples. They first used clay as a binder. Later, ancient Egyptians experimented with gypsum and lime to produce a mortar that could bind one stone to another. During the Roman Empire, the Colosseum and other ancient Roman structures were held together with a type of cement made of slaked lime and pozzolana. *Slaked lime* is lime heated and crumbled by adding water. *Pozzolana* is a volcanic ash from

Mount Vesuvius. This mixture was perhaps the first hydraulic cement. *Hydraulic cement* is a cement that hardens while submerged in water. The fact that these structures, whole or part, are still standing is evidence of the durability of the mortar used and of Roman engineering.

After the fall of the Roman Empire, the methods for making cement were lost. In the 18th century, John Smeaton, a British engineer, experimented with various mortar compositions. In 1756 he discovered that cement made from limestone with a substantial volume of clay hardens under water. This was a rediscovery of hydraulic cement. This naturally occurring cement, produced by burning a mixture of lime and clay, represents the reintroduction of hydraulic cement.

In 1824, Joseph Aspdin, a bricklayer from Leeds, England, invented what he called **Portland cement.** This is made from finely pulverized material consisting of compounds of lime, silica, alumina, and iron. The name was a reference to the Isle of Portland, which is located just off the British coast. The color of Aspdin's cement resembled the color of stone quarried on the Isle of Portland. Today, most people do not know where the name "Portland" came from and mistakenly believe it to be a manufacturer's name.

In the United States during the 19th century, the construction of a series of canals created a large demand for cement. Naturally occurring cement deposits were discovered during the 19th century in New York, Indiana, and Pennsylvania. Portland cement was not manufactured in the United States until the 1870s. Even the United States' most prolific inventor, Thomas A. Edison, contributed to the development of the Portland cement industry in the US. In 1902, Edison introduced the first long kilns for making Portland cement. His kilns, introduced at his Portland Cement Works in New Village, New Jersey, were 150 feet long, approximately twice the length of conventional kilns of that day. Today, kilns can exceed 500 feet in length.

COMPOSITION AND MANUFACTURE

Today, Portland cement is manufactured worldwide by many companies. It is an essential ingredient in concrete and mortar, which represent the world's most widely used building materials. The annual value of shipments of Portland cement is approximately 4.2 billion dollars. This current production rate is approximately 1.25 billion tons. If all Portland cement manufacturers were considered one entity, the industry would rank high in the Fortune 500. Portland cement manufacturing, however, remains a regional industry. The largest US cement company accounts for only 12.5% of capacity. Approximately 98% of all cement manufactured in the United States is Portland.

By definition, cement is a finely pulverized material consisting primarily of compounds of lime, silica, alumina, and iron. One pound of cement contains 150 billion grains that will pass through a sieve water cannot. The openings in the sieve are 1/25,000 of an inch.

Lime and silica constitute approximately 85% of cement. The common ingredients in cement are limestone, shells, and chalk or marl combined with shale, clay, slate or blast furnace slag, silica, sand, and iron ore.

Wet and Dry Manufacture

The manufacture of Portland cement requires 80 separate and continuous operations. The equipment for the process is large and heavy. The process also requires enormous heat and energy. Two different processes are currently used in the manufacture of Portland cement. These are wet process and dry process. The first step in both processes is to reduce the size of the limestone rock by crushing it. Primary crushers reduce rock as large as an office desk into 6″ pieces. A secondary crusher then further reduces the rock to 3″ and smaller, Figure 11-1.

During the *wet process*, the properly proportioned raw materials are ground with water, mixed, and fed into a large, rotary kiln. In the *dry process*, the raw materials are ground dry, properly proportioned, and fed into the kiln in a dry state, Figure 11-2.

Once fed into the kiln, the raw materials are heated to approximately 2700°F (1480°C), Figure 11-3. To save energy, some manufacturers preheat the raw materials prior to placing them in the kiln. The energy used to preheat the raw materials is often captured from the previous batch. The kilns can be as much as twelve feet in diameter and forty stories high. The kilns are constructed so their axis is slightly inclined from horizontal, Figure 11-4. The raw material is loaded into the higher end of the kiln and fed toward the lower end. At this end, an extremely hot flame transforms the mix into a clinker. *Clinkers* are marble-size pieces of the mix. The clinkers discharged from the kiln are red hot. They are cooled to handling temperature in a variety of coolers.

The cooled marble-size clinkers are ground into micro-size particles as small as 1/25,000 of an inch. During the process, a small amount of gypsum is added to the clinkers to control the setting time of the cement. The powder produced by this process is super-fine. It is conveyed to silos where it is stored until shipped to customers.

Figure 11-1.
These photographs show typical clinkers from a cement manufacturing process. A—Clinkers may be in various forms. B—Clinkers as seen in the process. (Portland Cement Association)

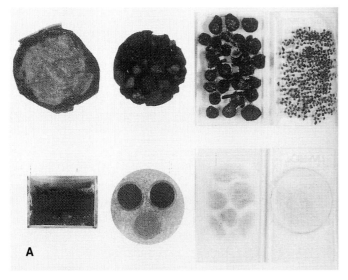

Figure 11-2.
In the dry process, raw materials are ground dry, properly proportioned, and fed into the kiln in a dry state. There are many possible proportions, each resulting in different properties.

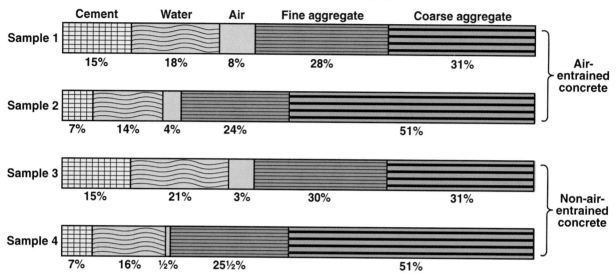

During the primary process, carbon dioxide in the raw materials is driven off. During this initial process, pollution control devices, electrostatic precipitators, or filters called *baghouses* remove particles from exit gases before they enter the atmosphere. Many manufacturing plants return all or a large portion of the collected particles to the kiln as part of the raw materials.

The majority of Portland cement is sold to customers in bulk delivered in trucks, railroad cars, or barges. Most of the bulk Portland cement is shipped to ready-mix plants that produce and sell concrete to local markets. A small percent of the Portland cement is bagged and sold to customers who need small amounts.

Portland cement manufactured in the United States conforms to the American Society for Testing and Materials (ASTM) C150 standard. This assures uniform quality among all Portland cement manufacturers.

Portland Types

Most of the Portland cement manufactured today is a general-purpose cement known as Type I. Other types of cement are produced for special applications by adding chemicals to the mix. Figure 11-5 shows the types of cement and their uses.

The "A" associated with Types I, II, and III Portland cement stands for *air-entrained.* Concrete made with air-entrained cements contain well-distributed air bubbles so small that there are billions of them in a cubic foot of concrete. Air bubbles in the concrete allow it to expand and contract without cracking. Sodium lauryl sulfate is the chemical commonly used for air entraining. Air entraining materials are incorporated in Portland cement by intergrinding them with the clinkers during the production process.

Figure 11-3.
This large rotary kiln is
used to make cement.
(Portland Cement Association)

Figure 11-4.
You can see in this photograph
the typical incline of a
rotary kiln.
(Portland Cement Association)

Figure 11-5.
This table shows the ASTM
types of Portland cement.

ASTM TYPES OF PORTLAND CEMENT	
Type	**Characteristics and Uses**
Type I, IA Standard	General purpose, most residential applications
Type II, IIA Modified	Reduced heat of hydration, increased sulfate resistance
Type III, IIIA High Early Strength	High strengths in 1–3 days, used in cold weather
Type IV, Low Heat	Heat during hydration kept to a minimum; intended for large masses, such as dams
Type V, Sulfate Resistant	Especially good for marine structures and in soils with high alkali content

One of the more significant developments in the manufacture of Portland cement was the introduction of air entraining. The air-entrained Portland cements were developed to produce concrete resistant to severe frost action and the effects of salt. As moisture in the concrete freezes and expands, the air bubbles act as expansion chambers.

THE CHEMISTRY OF CEMENT

The chemical process that causes cement to harden when mixed with water is called *hydration.* Equations (1) and (2) below are the hydration reactions of the two calcium silicates. This reaction between the Portland cement and water creates heat. The heat causes the paste formed by the mixture to begin to cure and harden. The principal product of the reaction of (1) and (2) is a calcium silicate hydrate. The heat generated by the reaction of the water with the Portland cement is proportional to the rate of reaction. The heat can be easily measured in the case of C_3S. C_3S is 40–70% by weight of the cement.

(1)
$$2\,C_3S + 7H \longrightarrow C_3S_2H_4 + 3CH$$
(Tricalcium Silicate)   (Water)   (C-S-H) + (Calcium Hydroxide)

(2)
$$2\,C_2S + 5H \longrightarrow C_3S_2H_4 + CH$$
(Dicalcium Silicate)   (Water)   (C-S-H) + (Calcium Hydroxide)

When first mixed, the reaction creates a rapid evolution of heat for approximately fifteen minutes. This is called Stage 1. The initial reaction is commonly referred to as *set* and occurs within 1 to 3 hours. Stage 1 is followed by a period of relative inactivity called Stage 2. The silicate continues to hydrate rapidly after the dormant period in Stage 2 at a rate corresponding to the maximum rate of heat evolution. This is Stage 3. Between approximately 2 and 3 hours, final set has passed and early hardening has started. At this point the hydration reaction has slowed. This is called Stage 4. The reaction continues to slow until it reaches a steady state, usually within 12 to 24 hours. This is Stage 5.

As with many chemical reactions, the hydration rate is affected by temperature. The rate of hydration increases as the temperature increases. Many people incorrectly think setting and hardening of cement is a product of evaporation. It is a product of hydration. The strength of the cement will increase as long as hydration occurs. This period can last for years.

CEMENT AND THE ENVIRONMENT

Generally, the cement industry has shown a commitment to managing its waste and emissions. The United States Department of Commerce reported that the cement industry invests 3.8% of the value of its shipments in pollution abatement. Manufacturing, overall, invests approximately 0.7% in pollution abatement. More efficient cement plants, resource recovery, and utilizing waste materials, such as used motor oil and scrap tires, have enabled the Portland cement companies to reduce their fossil fuel consumption by as much as 27% during the past 20 years. These changes help conserve the coal and natural gas that are used in great volumes to heat the kilns that produce the Portland cement. Energy is recycled during the manufacture of Portland cement as well. During the cooling of clinkers before grinding, heat is captured and used to preheat the raw materials before being placed in the kilns.

Progress has also been made in managing the solid waste generated during the manufacturing process. For example, cement dust from kilns is recycled back into the kilns, rather than being disposed of in a landfill. Waste materials from other manufacturing processes are also used as raw materials in cement making. Slag from steel making, fly ash, and used foundry sand are all sources of the silica, calcium, and iron needed for cement manufacture.

CONCRETE

Concrete is a monolithic ceramic product of aggregates bonded with cement. An *aggregate* is a reinforcement, such as sand, gravel, or slag. The cement is usually Portland, but can be asphalt or other cement.

Although concrete is conventionally classified as a ceramic, in reality it meets the definition of a composite as well. Concrete is a composition of fine and course aggregates in a matrix (cement). The cement coats the aggregates and binds the constituents together. The constituents in the mixture retain their identity, as they do in all composite materials.

Concrete by weight exceeds all other engineering materials used annually. It is used in the construction or manufacture of roads, bridges, culverts, drainage pipe, structural members, buildings, highways, and other construction projects. It is also used extensively for nonstructural architectural applications, such as curtain walls. See Figure 11-6.

Figure 11-6.
This building is constructed with concrete curtain walls. (Portland Cement Association)

One of the unique features of concrete is that it is a hydraulic material. A hydraulic material will cure or harden by interacting with water. Concrete, therefore, will cure while under water.

COMPOSITION

Concrete is a composite material consisting of large and small aggregates, cement, and water. The appropriate percentages of each constituent material is given below and illustrated in Figure 11-7.

- 11% Cement (usually Portland)
- 16% Water
- 6% Air
- 26% Sand
- 41% Gravel or Crushed Stone

For most structural work, the concrete ingredients are proportional to produce compressive strengths of 2500 psi to 5000 psi (16 MPa to 34 MPa). A *rich mixture* for columns may be in the proportion of 1 part cement, 1 part sand, and 3 parts stone. This is stated as a 1:1:3 ratio. A *lean mixture* for foundations may be a 1:3:6 ratio. This means 1 part cement, 3 parts sand, and 6 parts stone. Figure 11-8 shows some typical concrete mixtures.

CLASSIFICATION OF CONCRETE

Concrete is produced in four basic forms. These forms are ready mixed concrete, precast concrete, concrete masonry, and cement-based materials.

Ready Mixed Concrete

Ready mixed concrete is made off-site at a ready mix plant. It is delivered to a job site by a concrete truck. Ready mix concrete accounts for approximately 75% of all concrete made. Uses for ready mix concrete include footings, piers, walls, highways, and bridges.

Figure 11-7.
Several ingredients are used in different proportions to create concrete.

Figure 11-8.
This table shows various concrete mixes used in different applications.

CONCRETE MIXES FOR SELECTED APPLICATIONS				
Application	**Cement Content (Bags/Yd)**	**Max Size Aggregate**	**Water to Cement Ratio (Gals/Bag)**	**28th Day Compressive Strength**
8" Basement Wall	5	1–1 1/2"	7	2800
4" Basement Wall	6.2	1"	6	3500
Stairs	6.2	1"	6	3500
Driveways, Porches	6.2	1"	6	3500

Precast Concrete

In many applications, structural concrete components are cast at locations away from a job site. These precast concrete products are transported to the site by truck. Examples include bricks, bridge girders, wall panels, and culverts.

Concrete Masonry

Concrete masonry is cast off-site and includes a mixture of structural and architectural products. Uses include blocks, patio paving stones, and a wide variety of architectural units for residential and commercial construction.

Cement-Based Materials

Cement-based materials are manufactured off-site. Uses for these materials include mortar, grout, terrazzo, roofing tile, and countertops.

STANDARDIZED TESTS OF CONCRETE

There are a number of standard tests for concrete. However, three tests are commonly performed. These are ASTM C-143 Slump Test for Concrete, ASTM C-39 Compressive Strength Test, and ASTM C-78 Flexural Strength of Concrete Test.

ASTM C-143 Slump Test for Concrete

Slump is a measure of how much new concrete "falls" when removed from a frame. In other words, it is how viscous the wet concrete is. The *slump test* is used to measure the consistency of concrete. The degree of slump is a function of the mix. The amount of slump, for example, may depend on the amount of water or the type and size of aggregate.

The slump test is performed by placing newly mixed concrete in a slump cone. The slump cone is placed on a flat surface, usually a plank or slab of concrete. The dimensions of the slump cone are shown in Figure 11-9.

The procedure begins by filling the slump cone 1/3 full of the concrete mix and puddling it 25 times with a 5/8" bullet-pointed, 24" long metal rod. *Puddling* is stroking up and down, and done to fill all voids in the slump cone. The next step is to fill the cone to 2/3 full and puddle it again 25 times. Finally, the cone is filled completely and puddled a final 25 strokes. The excess is then removed from the top of the cone by using a straight metal rod and carefully working back and forth across the top of the metal cone to level the concrete. This process is referred to as *striking off* the concrete.

Once the cone is filled and the excess struck off, the slump cone is carefully lifted from the wet concrete and set beside it. The slump is measured by laying a straight edge across the top of the cone and over the concrete. The distance from the top of the cone to the top of the concrete is the slump. The slump should remain consistent for a single job with each new mix of concrete.

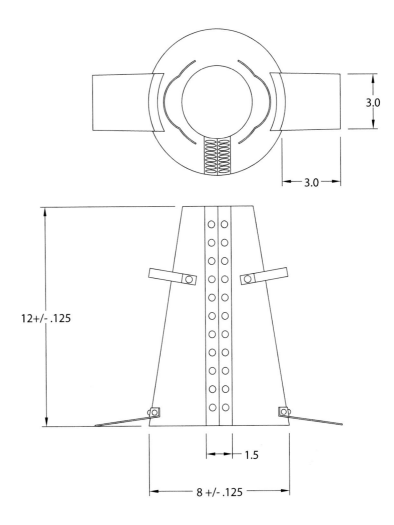

ASTM C-39 Compressive Strength Test

As compressive strength is perhaps the most important property of concrete, the compressive strength test is performed frequently. The compressive strength is performed by placing newly mixed concrete in a cylinder of ASTM standard height and diameter. A 6" diameter and 12" high cylinder is generally used. However, the standard states the cylinder is to have a diameter-to-height ratio of 1 to 2 (ASTM C-192).

According to the ASTM standard C-39, neither end of the specimen shall depart from perpendicularity to the axis by more than 0.5° (approximately 1/8" in 12"). If the specimen does not conform, it may be cut or ground to meet the specification. The length to diameter ratio, however, cannot be less than 1.8.

The process begins by placing the mixed concrete into the cylinder. This is accomplished by first filling the cylinder 1/3 full and puddling it 25 times with a 5/8" diameter, bullet-pointed steel rod. This step is repeated two more times to completely fill the cylinder. Any excess is struck off with a trowel and the cylinder is capped until tested. The samples are allowed to cure in the capped cylinders for 1, 3, 7, and 28 days before testing. Figure 11-10 shows the time tolerance allowed by ASTM.

The testing procedure involves applying a universal load on each specimen until it fails. To calculate the compressive strength the cross sectional area of the cylinder and the applied load are needed. The formula is:

$$\text{Compressive Strength } \sigma = \frac{\text{Load}}{\text{Cross Sectional Area (CSA)}}$$

For example, the load at failure is 98,000 lbs. and the specimen diameter is 6″. The cross sectional area is πr^2, or 28.27 in^2. Therefore,

$$\text{Compressive Strength } (\alpha) = \frac{98,000 \text{ lbs}}{28.27 \text{ in}^2} = 3466.57 \text{ psi}$$

Compressive strength is affected by curing time and moisture content, Figure 11-11. The strength increases as the curing time increases and the water decreases.

ASTM C-78 Flexural Strength of Concrete Test

Although compressive strength is generally of greater interest to concrete users, there are certain applications where tensile strength is required. This is especially true when concrete is placed in a horizontal plane, such as in bridges, floors, and roofs.

A true measure of tensile strength of concrete is difficult to obtain. One reason is the difficulty in gripping the specimen in order to apply a true tension force. The bending test provides a measure of modulus of rupture (MOR). The MOR is closely related to tensile strength. ASTM C-78 Standard Test Method for Flexural Strength of Concrete (using Simple Beam with Third Point Loading) is the standard test for determining the flexural strength of concrete.

The testing method for flexural strength is illustrated in Figure 11-12. The test specimen must have a span three times its depth, within 2%. The sides of the specimen must be at right angles to the top and bottom surfaces.

Figure 11-10.
This table shows the time tolerances for a concrete compressive strength test as outlined by ASTM.

ASTM TIME TOLERANCES FOR CONCRETE COMPRESSIVE STRENGTH TEST	
Test Age	**Permissible Tolerance**
24 Hours	±0.5 Hours or 2.1%
3 Days	2 Hours or 2.8%
7 Days	6 Hours or 3.6%
28 Days	20 Hours or 3.0%
90 Days	2 Days or 2.2%

Figure 11-11.
These graphs show the water content and compressive strength of concrete over time.

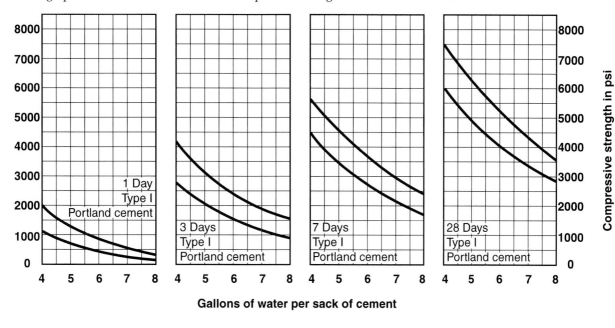

Gallons of water per sack of cement

Figure 11-12.
ASTM standards describe how to conduct a flexural test of concrete.

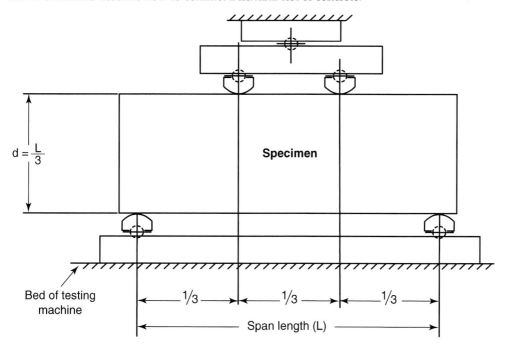

All surfaces that are in contact with the load and support blocks must be smooth and free of scars, indentations, holes, or inscribed identifications. The load on the specimen may be applied rapidly up to 50% of the assumed breaking load. The load must then be reduced to 125–175 psi (861–1207 KPa). This load rate is held until failure occurs. When failure occurs, the maximum load in pounds or kilograms is recorded and used to calculate the modulus of rupture.

$$MOR = Pl / bd^2$$

Where P is the maximum load, l is the span length, b is the average width of the specimen, and d is the average depth of the specimen. The average width and depth of the specimen is determined by taking measurements at each edge and at the center. The measurements are read to the nearest .05″ (1.3 mm) and averaged.

FIBER REINFORCED CONCRETE

Concrete has been reinforced in one way or another for many years. This has been done to overcome its brittle nature and low tensile strength. Traditionally, the reinforcing material has been either steel wires or solid round steel bar stock. The solid bar stock is commonly called *rebar* (reinforcing bar). Although steel is still very popular, especially for vertical applications, fibers are now being used as reinforcing materials as well. These fibers are made of a variety of materials including glass, Kevlar, polypropylene, nylon, and cellulose.

Adding fibers not only strengthens the concrete, but also improves its ductility. When a crack occurs in a brittle material, the crack tends to grow. This is commonly referred to as *crack propagation.* The fibers used to reinforce concrete are short and not continuous. They are placed randomly in the concrete mix. These fibers tend to stop a crack from propagating.

Fiber reinforced concrete is grouped into three types.
- Low Fiber Volume Composite
- Higher Fiber Volume Composite
- High Fiber Volume Composite

Low fiber volume composite concrete contains less than 1% fiber. It is used for field applications involving large volumes of concrete. The fibers do not significantly increase the strength of concrete. The fibers are used to control cracking. Low fiber volume composite concrete is used for paving roads.

Higher fiber volume composite concrete is typically used for thin sheets with cement mortar mix. The fiber volume in this mix ranges from 5% to 15%.

High fiber volume composite concrete has the greatest volume of fiber reinforcement. This significantly increases its strength and toughness. The reinforcement in high volume fiber composite concrete is usually a sheet material. This concrete is used in roof and wall panels, especially for commercial structures. The fiber volume in this mix can be as high as 40%.

SUMMARY

Cement and concrete are among the oldest building materials. Early Egyptian and Roman civilizations used materials similar to today's cement and concrete to construct tombs, temples, and fortifications. Some of these buildings have lasted for centuries. The Romans even developed a form of hydraulic cement. This is a cement that hardens while submerged in water.

The cement used today is most often Portland cement. Portland cement was introduced during the 18th century by John Smeaton, a British engineer. It is a finely pulverized material consisting principally of compounds of lime, silica, alumina, and iron. It is manufactured in large quantities worldwide. Portland cement is an essential ingredient in concrete and mortar.

Concrete is truly a ceramic matrix composite composed of small and large aggregates held together by cement. Concrete is cured by a chemical process known as hydration. Concrete is a hard brittle ceramic composite used for many applications. Its compressive strength makes it especially good for footings, piers, and foundations. When reinforced with steel or fiber, concrete is also used for floors, walls, bridges, and nonstructural architectural applications.

IMPORTANT TERMS

Aggregate	Low Fiber Volume Composite
Air-Entrained	Portland Cement
Baghouses	Pozzolana
Cement	Puddling
Clinker	Rebar
Concrete	Rich Mixture
Crack Propagation	Set
Dry Process	Slaked Lime
High Fiber Volume Composite	Slump
Higher Fiber Volume Composite	Slump Test
Hydration	Striking Off
Hydraulic Cement	Wet Process
Lean Mixture	

QUESTIONS FOR REVIEW AND DISCUSSION

1. Define concrete.
2. What is the difference between cement and concrete?
3. What early civilizations used some form of cement and concrete?
4. What is pozzolana? What is its significance?
5. What term is used to describe a ceramic material that can harden under water?
6. Who is credited with inventing Portland cement? How did the name Portland cement originate?
7. What event created the first large scale demand for cement in the United States?

8. What famous United States inventor contributed to the development of the Portland cement industry? What contribution did he make?

9. What are the primary ingredients in Portland cement?

10. Describe some of the ways the cement industry has become environmentally active in their manufacturing process.

11. To what industry is the majority of the manufactured Portland cement sold?

12. What significance does the letter "A" have in Type IA Portland cement?

13. Explain the chemical process of hydration.

14. For most structural applications, concrete is proportioned to attain compressive strengths of _____.

15. List three products made of precast concrete.

16. What property of concrete does the slump test determine?

17. What purpose does puddling serve in preparing samples for slump and compressive strength tests of concrete?

18. Describe how the slump of concrete is measured.

19. A sample of concrete was taken from a highway project to be tested for compressive strength. The sample when cured measured 5.88" in diameter and 11.98" in height. When tested after seven days for compressive strength, it fractured at a load of 89,000 lbs. Calculate its compressive strength.

20. Determine the modulus of rupture for a concrete specimen whose dimensions are $3'' \times 3'' \times 11''$. The span between bearing points is 9", and the load at fracture was 9,500 lbs.

21. What materials have been traditionally used to reinforce concrete?

22. Other than providing additional strength, what function do the fibers in fiber reinforced concrete serve?

23. List the three types of fiber reinforced concrete.

24. What materials are used as reinforcing fibers in fiber reinforced concrete?

FURTHER READINGS

1. Borsum, Michael. Fundamentals of Ceramics. New York: McGraw-Hill Inc. (1997).

2. Brady, George S. and Clauser, Henry R. Materials Handbook 13th ed. New York: McGraw-Hill, Inc. (1991).

INTERNET RESOURCES

www.portcement.org
Portland Cement Association

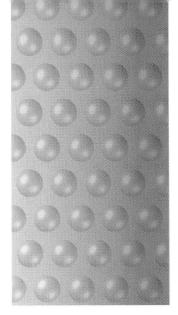

12

Introduction to Composites

KEY CONCEPTS

Upon completion of this chapter, you should understand:
- ➤ Definitions and basic structure of composites.
- ➤ Polymer, metal, and ceramic matrices.
- ➤ Use of fiber, whisker, and particulate reinforcement.
- ➤ Interphase/interface.
- ➤ The rule of mixture.
- ➤ Design and fabrication of composites.
- ➤ Standards for composites.

New requirements for high performance defense and space systems have led to the development of advanced engineered materials. Service requirements that could not be met with conventional materials are now being fulfilled with advanced composites and ceramics. Recently developed composites are able to retain strength at high temperatures and are often lighter in weight than conventional materials. More than 10,000 pounds of advanced composites are used in each space shuttle. They are also used extensively in the construction of the international space station.

The value of components produced from advanced materials in 1988 was approximately 2 billion dollars. This figure increased to 20 billion dollars by the year 2000. Most of the advanced ceramics and composites are currently being used by governments in space or advanced weapon system projects. In commercial use, the aircraft and sporting goods manufacturers use most of these materials. This is predicted to change as the materials become less expensive and manufacturing processes for high-volume production improve.

The development of composites is also changing the design process. Traditionally, engineers select materials that most closely meet specific service requirements. In many cases, a single material could not satisfactorily meet the requirements. Now, properties can be designed into the materials to meet special service requirements of high performance products. As a result, composites are often referred to as "tailored" or "tailorable" materials.

Although advanced composites and ceramics are used extensively in the aerospace industry, they are rapidly becoming a primary manufacturing material in the automotive, machinery, sporting goods, construction, and commercial aircraft industries. The sporting goods industry has taken the lead in the commercial market. Advanced composites have been used to manufacture a variety of sporting goods including golf clubs, fishing rods, tennis rackets, boats, and track and field equipment.

Research is being conducted around the world to develop composites that operate at higher temperatures and lower costs. The United States government currently spends about $170 million a year for research and development related to advanced ceramics and composites, Figure 12-1. This is more than any other nation. Japan, however, has been very aggressive in developing and using advanced ceramics and composites in commercial products.

DEFINITIONS AND ADVANTAGES

Composites consist of two or more materials or material phases combined to produce a material with superior properties to those of its individual constituents. Although the focus of this chapter is on composites of a more advanced nature, it is important to understand that they have been used for centuries and that composites exist in nature. Wood is perhaps the best example of a natural composite. Wood consists of fibers made of cellulose in a matrix of lignin, Figure 12-2. Some composites manufactured by humans

Figure 12-1.
This table shows government funding for advanced materials.
(US Congress, Office of Technology Assessment)

US GOVERNMENT AGENCY FUNDING FOR ADVANCED STRUCTURAL MATERIALS Fiscal Year 1987 (millions of dollars)					
Agency	Ceramics and ceramic matrix composites	Polymer matrix composites	Metal matrix composites	Carbon/carbon composites	Total
Department of Defense	$21.5	$33.8	$29.7	$13.2	$ 98.2
Department of Energy	36.0	—	—	—	36.0
NASA	7.0	5.0	5.6	2.1	19.7
National Science Foundation	3.7	3.0	—	—	6.7
National Bureau of Standards	3.0	0.5	1.0	—	4.5
Bureau of Mines	2.0	—	—	—	2.0
Department of Transportation	—	0.2	—	—	0.2
Total	$73.2	$42.5	$36.3	$15.3	$167.3

have been used for thousands of years, such as brick. Plaster, concrete, plywood, and flakeboard are a few examples of modern materials classified as composites.

Composites are based on the controlled distribution of one or more materials in a continuous phase of another. All composites consist of a reinforcement, matrix, and a boundary between. This boundary is known as the interphase, Figure 12-3. The *reinforcement* is that phase of the composite providing strength. The reinforcement is generally in the shape of fiber, whisker, or particulate. The *matrix* is the glue that holds the reinforcement together. For example, concrete is a composite consisting of sand and stone reinforcement in a matrix of cement.

Figure 12-2.
The microstructure of wood includes lignin and cellulose.

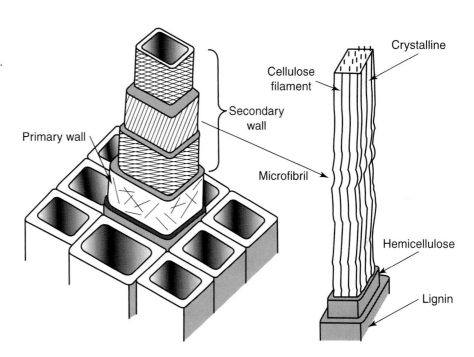

Figure 12-3.
A composite includes a matrix, reinforcement, and interphase.

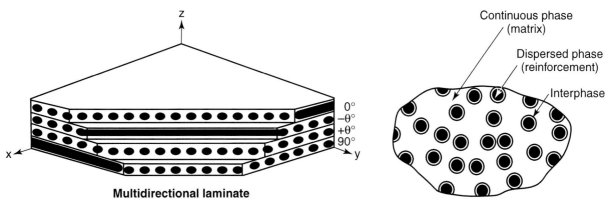

The reinforcement, matrix, and interface contribute to unique properties over conventional materials. Generally, the use of composites has advantages over conventional materials. The following list illustrates a few examples.

- **Design flexibility:** The properties of composites are dependent on the constituent parts. Therefore, the properties can be tailored to meet different design requirements by varying the composite makeup.
- **Weight reduction:** Weight savings are significant when composites are used. For example, unidirectional aramid and carbon fiber reinforced epoxies provide a specific tensile strength that is four to six times greater than that of steel or aluminum. Yet, the material is 25% to 50% lighter than the metals.
- **High torsional stiffness:** The high torsional stiffness requirements of various vehicles, particularly high-speed aircraft, can be satisfied with composites.
- **Outstanding corrosion and fatigue resistance:** Both corrosion and fatigue in aluminum alloys present problems costly to remedy. In response to these problems, fiber-reinforced composites have been used successfully in filament-wound rocket motors and in various other structural applications.
- **Simplified production processes:** Manufacturing and assembly are simplified because of joint and fastener integration. This can reduce engineering, purchasing, and follow-up costs.

RULE OF MIXTURE

The properties of a composite depend directly on the proportions of the constituent materials. For example, a 50%-50% mixture of fiber and polymer has a density midway between those of the two components. Expressed as an equation, the density of the composite P_c is

$$P_c = (1-V_f)P_m + V_f P_f$$

where V_f is the volume fraction of fibers in the composite, P_m is the density of the matrix, and P_f is the density of the fibers. The equation shows that the density of the composite is proportional to the densities of the fibers and matrix. This is an example of the *rule of mixtures* for composites. This rule is used frequently to predict and design a range of other composite properties.

However, in reality, not all of the properties of a composite are in simple proportion to those of its constituents. Sometimes the properties of a composite are far superior to those of its individual components. As a result, the material as a whole is better than the sum of the parts.

REINFORCEMENT, MATRICES, AND INTERPHASE

As stated previously, a composite consists of two or more materials or material phases combined to produce a material with superior properties to those of its individual constituents. Every composite, natural or synthetic, has three phases—reinforcement, matrix, and interphase. The selection of reinforcement, ratio of reinforcement to matrix, orientation of reinforcement, and other factors determine the properties of the composite.

REINFORCEMENT

Reinforcement is primarily responsible for the structural properties of a composite. Structural properties include strength and stiffness. Materials used as composite reinforcements fall into one of three categories.

- Fibers
- Whiskers
- Particulates (Flake and Sphere)

Fibers

Fibers have been the primary reinforcements used for composites since the 1950s. Most structural composites are reinforced with glass fibers. These products are called *fiberglass* and were first commercialized in the late 1930s. Currently, fiberglass accounts for approximately 90% of the reinforced plastics market. The light weight, high strength, and nonmetallic characteristics of fiberglass products led to wide acceptance during World War II. Boat hulls, corrugated roofing, fishing rods, and a wide range of other products are manufactured from fiberglass, Figure 12-4.

The glass fibers commonly used as reinforcements are E glass and S glass. Other fiber reinforcement materials include carbon/graphite (CG), boron, and aramid. Carbon/graphite (CG) and aramid fibers are often used in the aerospace industry. Figure 12-5 compares various properties of selected fiber reinforcement materials used in advanced composites.

Glass fibers

E glass, or electrical glass, is calcium aluminoborosilicate. It provides excellent dielectric properties, chemical resistance, corrosion resistance, and fatigue resistance. It also has a high strength-to-weight ratio. E glass retains 50% of its tensile strength up to 650°F (345°C). E glass is available as continuous filament, chopped fiber, and fiber cloth.

Figure 12-4.
Many new homes have roof shingles made from fiberglass.

Figure 12-5.
This table shows the properties of fiber reinforcements for advanced composites.

PROPERTIES OF FIBER REINFORCEMENTS FOR ADVANCED COMPOSITES						
Property	**Fiber**					
	Carbon/Boron	Aramid	E Glass	S Glass	Graphite	PEEK
Tensile Strength (10^3 psi)	510	400	500	660	300	75
Young's Modulus (10^6 psi)	58	18	10.5	12.5	32	—
Elongation (%)	19	2.4	4.8	5.4	1.2	20
Coef. of Therm Exp. (10^{-6} °F in/in) Long	2.5	−1.1	1.6	2.8	−0.55	—
Specific Gravity	2.58	1.44	2.54	2.49	1.8	1.3

S glass, or high-strength glass, is soda-lime silica. It is used for components that require higher strengths. S glass has higher compressive and tensile strength, a higher modulus of elasticity, and a slightly lower density than E glass. It is also significantly more expensive.

Carbon/graphite fibers

Carbon/graphite (CG) fibers are the most frequently used reinforcement material for composites used in advanced systems. CG fibers are produced from organic precursor fibers such as polyacrylonitrile (PAN) or by means of pyrolysis. An *organic precursor fiber* is an intermediate step in producing a final substance. *Pyrolysis* is burning of a polymer, such as rayon, in a high-temperature furnace. The product of this process is a strong, stiff fiber a few microns in diameter.

By controlling the process, the tensile strength and stiffness can be raised or lowered for different applications. CG fibers are generally produced with intermediate strength and stiffness for aircraft applications. Spacecraft applications generally require fibers with high stiffness. High strength fibers are used in missile applications. Graphite epoxy composites are used for a wide range of aerospace structures, both current and proposed.

Boron fibers

Boron fibers have been used extensively in the aerospace industry. First introduced in 1959, their relatively high cost has kept them from being used in high volume applications. Boron fibers have excellent compressive strength and stiffness. They are also extremely hard. Boron fibers are manufactured by passing boron vapor over a thin filament of tungsten. The boron is deposited on the tungsten filament.

Aramid fibers

Aramid fibers are organic fibers introduced in the early 1970s by DuPont. Commercially known as Kevlar®, this fiber has been used in a wide variety of products including helmets, jet fighters, and commercial aircraft. Kevlar® is also used as a reinforcement fiber in high-performance motorcycle helmets and bullet-proof vests, Figure 12-6. DuPont's Nomex®, Teijin's HM50®, and Sumitomo Chemical's Ekonol® are examples of other aramid fibers on the market.

Approximately 2500 pounds of Kevlar is used in the fabrication of the Lockheed L1011-500 aircraft. Fairings, panels, and control surface components are fabricated from Kevlar 49 epoxy laminate. Kevlar 49 has excellent electrical properties, high tensile strength and toughness. However, it has less compressive strength compared to carbon/graphite.

Whisker

Ceramics *whiskers* are the most common whisker used as reinforcement in advanced composites. Whiskers have smaller aspect ratios than fibers. Whiskers are very fine, single crystal fibers that range from 3 to 10 micrometers in diameter. Although brittle by nature, ceramic whiskers retain their strength at high temperatures. Ceramic whiskers are generally single crystals with an aspect ratio of 50 or more. The aspect ratio is the length divided by the diameter. Silicon carbide (SiC), silicon nitride (Si_3N_4), and high purity alumina, are common types of whisker reinforcement materials. Silicon carbide is the most common whisker reinforcement.

Figure 12-6.
This bullet-proof vest has a lining reinforced with Kevlar® fibers.

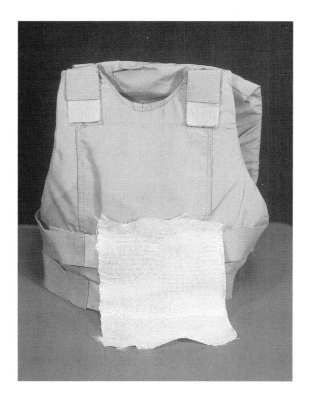

Whisker-reinforced ceramic composites have excellent insulating properties, high strength-to-weight ratio, and a low coefficient of thermal expansion. Their thermal stability has made them the material of choice in many electronic, military, and aerospace applications where high temperatures are encountered.

Particulates

Glass, ceramics, and metal are used to form *particulates* for reinforcement. Particulate is a classification of reinforcement characterized by a very low aspect ratio. Often the ratio is 1:1. For example, the stone in concrete is classified as particulate. They are used to reinforce metal, plastic, and ceramic matrix composites. Particulate reinforced composites are anisotropic. This means they have different properties depending on the direction of measurement. Cutting tools, seals, bearings, valves, heat engines, and medical implants are a few of the uses for particulate reinforced composites. In addition, some particulates can provide excellent electrical properties for the composite.

Flake reinforcement is used primarily when uniform properties are required. Ceramic flake reinforced composites make excellent insulators. Metal flake reinforced composites make good conductors.

Sphere reinforcements aid in resisting shock and have excellent compressive strength. They are dimensionally stable and exhibit low coefficient of thermal expansion. These properties make them extremely useful for a wide range of products.

MATRIX

The *matrix* is the "glue" that bonds fibers, whiskers, or particulates together to form one material. However, the matrix also transfers loads between the particles of reinforcing material. Protecting the reinforcement is yet another function of the matrix. The matrix also protects the reinforcement against abrasion and corrosion. The corrosion can be caused by the environment, such as moisture, oxidation, and chemical corrosion.

Composites are generally classified by the type of matrix used in their production. The matrix can be polymer, metal, or ceramic. Therefore, composites are classified as polymer matrix composite (PMC), metal matrix composite (MMC), and ceramic matrix composite (CMC).

Polymer matrices

Polymer matrices (plastic resins) have dominated the composite industry since World War II. Both thermoset and thermoplastic polymer resins are used as the matrix in the composite. Until recently, only thermosets were primarily used. Thermoplastics were considered to have inferior properties when compared to thermosets. Recent advances, however, in thermoplastic technology have led to the development of several new resins that compare favorably to the thermosets.

Thermosetting resins tend to have better heat and chemical resistance and greater dimensional stability than thermoplastics. This is a result of the structural difference between thermosetting and thermoplastic resins. During the curing process, the polymer chains in thermosetting plastics cross-link to form a three-dimensional network, Figure 12-7. Thermoplastics do not cross-link and, therefore, are not as resistant to heat and chemicals. Figure 12-8 compares general characteristics of thermoset and thermoplastic matrices.

Epoxies are the most used thermosetting resin. Epoxies are high strength adhesives combined with a variety of reinforcements. The result is high strength materials used in the aerospace and other industries. When combined with graphite they produce a strong, tough, dimensionally stable material. The service temperature limit for epoxies generally does not exceed 350°F. However, specialty epoxies are commercially available with temperature limits to 475°F (245°C). Epoxies are relatively expensive when compared to other resin systems.

Other commonly used thermoset resins include, polyimides, vinylesters, bismaleimides, and some polyesters. Polyesters are used for high volume production of auto parts and for large marine structures.

Figure 12-7.
Thermosetting plastics can have cross-linked polymer chains.

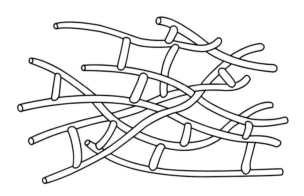

Figure 12-8.
This table shows general characteristics of thermoset and thermoplastic matrices. (US Congress, Office of Technology Assessment)

GENERAL CHARACTERISTICS OF THERMOSET AND THERMOPLASTIC MATRICES					
Resin Type	**Process Temperature**	**Process Time**	**Use Temperature**	**Solvent Resistance**	**Toughness**
Thermoset	Low	High	High	High	Low
Toughened thermoset	↑	↑	↑	↑	↑
Lightly cross-linked thermoplastic	↓	↓	↓	↓	↓
Thermoplastic	High	Low	Low	Low	High

The fact that thermoplastics do not cross-link can be an advantage or a disadvantage. An advantage is thermoplastics can be processed faster than thermosets. This is a great advantage to high-volume industries, such as the automotive industry. The composites made with thermoplastic resins tend to be tougher and more resistant to cracking and impact damage. Also, thermoplastics can be formed a number of times if the initial forming process fails. Thermosets cannot be recycled nor can they be formed more than one time. The biggest disadvantages of thermoplastics are low heat and chemical resistance.

Thermoplastic resins include polyetherimide (PEI), polyimide, polyphenylene sulfide (PPS), thermoplastic polyimide (TPI), polyetheretherketone (PEEK), and some polyesters. PEEK and some of the other newly developed thermoplastics compare favorably with thermosets in high temperature strength and chemical resistance. These resins are generally used with glass, graphite, boron, and aramid fibers.

Metal matrices

Metal matrices (metallic powders) in composites generally include low density metals, such as aluminum or magnesium. Metal matrix composites have high strength, high stiffness, and excellent dimensional and thermal stability. When compared to conventional metals and plastic matrix composites, they are also resistant to environmental factors.

Compared to conventional, unreinforced metals, metal matrix composites have significantly greater strength, Figure 12-9. However, they have much lower ductility and toughness. The fracture toughness of metal matrix composites is generally lower than that of conventional unreinforced metals. This is a trade-off for their higher strength and stiffness.

Fiber reinforced metal matrix composites are excellent conductors of heat and electricity. The heat or electrical current is transmitted from fiber to fiber much more efficiently because of the high aspect ratios of fiber

Figure 12-9.
This table shows a comparison of conventional metals to mmc.

COMPARISION OF CONVENTIONAL METALS TO METAL MATRIX COMPOSITES (MMCs)							
Property	**Aluminum 6061**	**Titanium Ti-6Al-4v**	**Graphite/ Aluminum**	**Boron/ Aluminum**	**Silicon Carbide Alumina**	**Alumina Aluminum**	**Silicon Carbide Titanium**
Tensile Strength (axial) MPa	290	1170	690	1240	1040	620	1720
Tensile Strength (transverse) MPa	290	1170	30	140	70	170	340
Stiffness (axial) GPa	70	114	450	205	130	205	260
Stiffness (transverse) GPa	114	114	34	140	99	140	173

reinforced metal matrix composites. Graphite fibers in a matrix of aluminum or copper are especially good thermal conductors.

A major disadvantage of metal matrix composites are costs associated with the constituent materials and the processing used to produce the composite structure. Generally, these costs restrict their use to high performance military and aerospace applications. Applications include spacecraft structures and military aircraft engines.

Although commercial use of metal matrix composites is limited, there are some applications. An aluminum piston selectively reinforced with a ceramic fiber has been developed for a diesel truck engine, Figure 12-10. Other commercial applications include golf clubs, tennis rackets, skis, and other sporting goods.

Ceramic matrices

A wide range of applications for ceramic matrix composites exist. These include cutting tools, bearings, seals, valves, orthopedic implants, heat exchangers, heat engines, diesel engines, and gun barrel liners.

The most commonly used *ceramic matrix* (ceramic slurry) is cement. Portland cement, used as a binder in concrete, is a ceramic matrix that was used long before "advanced composites" were even theorized. Advanced ceramic composites consist of glass or ceramic matrices reinforced with fibers, whiskers, or particulates.

Perhaps the most popular glass/ceramic matrices are lithiumalumina-silicate (Li_2O-Al_2O_3-SiO), commonly referred to as LAS; BMAS (BaO-MgO-Al_2O_3-SiO); and magnesia-alumina-silicate (MgO-Al_2O_3-SiO). These matrices and others are typically reinforced with fibers, whiskers, or particulates of silicon carbide (SiC), silicon nitride (Si_3N_4), zirconium oxide (ZrO_2), aluminum oxide (Al_2O_3) or other engineered ceramics.

Figure 12-10.
This truck piston is selectively reinforced with ceramic fiber.

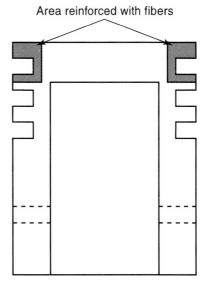

Area reinforced with fibers

Research in ceramic matrix composites generally focuses on ways to increase toughness. Ceramics are, by nature, very brittle. They do not withstand impact or shock forces very well. Researchers, however, have been able to increase fracture toughness in ceramic composites over monolithic ceramic structures.

Generally, ceramics have high strength at high temperatures, low thermal conductivity, and are extremely hard. By reducing the strength of the bond between the matrix and the reinforcement, the fracture toughness of ceramic composites can be increased. Long fibers in the ceramic or glass matrix can increase the toughness by a factor of two over monolithic ceramics. This is because fibers stop the propagation of cracks caused by shock or impact. Figure 12-11 compares physical and mechanical properties of various ceramics and ceramic matrix composites.

Compatibility between the matrix and the reinforcement is especially important in ceramic composites. For example, look at the coefficient of thermal expansion. If the coefficient of the matrix and the reinforcement are significantly different, the component may fail on temperature change. Compatibility is generally achieved when the matrix and reinforcement are from the same material.

INTERPHASE

The boundary between the matrix and the reinforcement in a composite is known as the *interphase* or *interface*. The interphase generally does not receive the same attention as the matrix and reinforcement. However, it

Figure 12-11.
This table shows the properties of various ceramic matrix composites.

PROPERTIES OF SELECTED CERAMIC MATRIX COMPOSITES					
Property	Alumina 99.5	SiC Reaction Bonded	Si_3N_4 Reaction Bonded	Al_2O_3	ZrO_2
Compressive Strength psi $\times 10^3$	380	100	112	350	300
Tensile Strength psi $\times 10^3$	30	20	—	—	21
Flexural Strength psi $\times 10^3$	50	37	30		100
Modulus of Elasticity psi $\times 100$	55	56	24	38	30
Impact Resistance in./lbs charpy	6	<.8	<.8	—	—
Hardness MOH´s	9	9	9		
Knoop				1470	1100
Density Kg/$M^3 \times 10^{-3}$	3.5	3.21	3.44		4

is an important aspect when engineering composite structures. The bond between matrix and reinforcement affects the overall performance of the composite. The interphase is especially important in controlling the mechanical behavior of composites.

The most common way to control the mechanical behavior is to increase the strength of the bond between the matrix and reinforcement. However, there are composites that perform better with a weaker bond. For example, the toughness of ceramics is increased if the bond between the matrix and reinforcement is weaker. Polymer reinforcements, on the other hand, are often treated with a coupling agent to increase the bond between the constituents of the composite component.

Coupling agents are used to improve the interface between the matrix and the reinforcement. They can perform one of several functions. For example, coupling agents can improve wetability. *Wetability* is the ability of any solid to be enveloped when in contact with a liquid. This increases the bond strength between the reinforcement and the matrix.

PROPERTIES OF COMPOSITES

Instead of attempting to describe the properties of the nearly infinite possible polymer, ceramic, or metal matrix composites, this text looks at the properties of the elements that make up the composites. The properties of a composite is determined by the elements it is made from. For example, an aluminum or copper matrix with long carbon fibers conducts electricity and heat much more efficiently than an aluminum matrix with carbon flakes. The length of the carbon fibers, in this case, affect the property of the composite. The following is a partial list of the factors that affect the properties of composites.

- ◆ Composition of Components
- ◆ Size of Components
- ◆ Shape of Components
- ◆ Proportion of Reinforcement to Matrix
- ◆ Orientation of Reinforcement
- ◆ Interphase Adhesion

COMPOSITION OF COMPONENTS

Several generalizations can be made about the properties of polymer, metal, and ceramic matrix composites. Polymer matrix composites are lighter in weight and tougher than either metal or ceramic composites, but their service temperature is generally much lower. Polymers melt or char at temperatures above 600°F (315°C). Some metals generally can retain their strength to about 1900°F (1040°C). Ceramics can retain their strength to well above 3000°F (1650°C).

Metal matrix composites are good conductors of heat and electricity. They are much better conductors than either polymer or ceramic matrix composites.

Ceramic composites are brittle but high in compressive strength and have a low coefficient of thermal expansion. Figure 12-12 compares specific strength and stiffness of some advanced materials to conventional metals.

Wear resistance is much higher in metal matrix and ceramic matrix composites than it is in polymer matrix composites and in unreinforced alloys or ceramics. The alumina-silica reinforced aluminum truck piston described earlier had an 85% improvement in wear resistance over the conventional type piston.

Thermal expansion can be reduced to near zero by choosing the proper matrix and reinforcement. Silicon carbide particulates in aluminum result in a much lower coefficient of thermal expansion when compared to the conventional metals.

SIZE AND SHAPE OF COMPONENTS

Preformed composites are often classified by the shape of the reinforcement. These are fiber, whisker, or particulate, Figure 12-13. Each type of reinforcement adds unique properties to the composite.

Figure 12-12.
This chart compares the strength of conventional materials to composites.

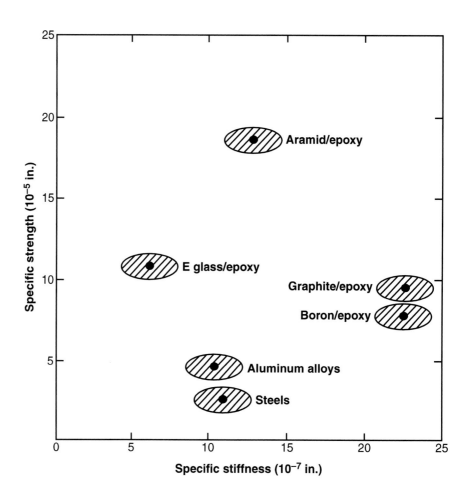

Figure 12-13.
Reinforcements can be in the form of fiber, whisker, or particulate.

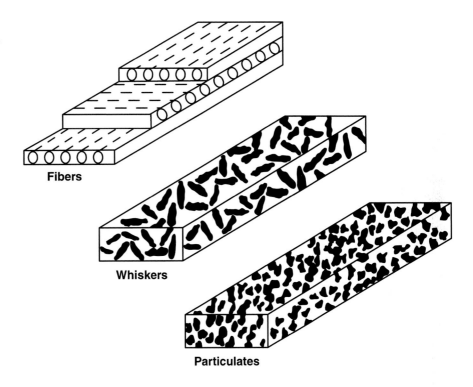

Fibers

Whiskers

Particulates

The aspect ratio of the reinforcement is the most common criteria used to describe the physical dimensions of reinforcements. It is important in determining the properties of the composite. *Aspect ratio* is the ratio between the diameter and the length of the reinforcing material. It is calculated by dividing the length of the individual reinforcement by its diameter. Aspect ratios can easily exceed 500 for fibers or be as little as 10 for particulates.

Longer, continuous fibers in polymer matrix composites increase their strength. Longer fibers, or high aspect ratio fibers, in metal matrix composites improve electrical and thermal conductivity. An aspect ratio of 300 in E glass fibers increases the tensile modulus by 65% over shorter fiber reinforcements with an aspect ratio of 50. Longer fibers in ceramic matrix composites help increase the toughness of ceramics. The longer fibers tend to stop the propagation of cracks caused by shock or impact.

Reinforcements with low aspect ratios, typically particulates, are also used to design specific properties into composites. Spheres and flakes generally improve the compressive strength, shock resistance, and dimensional stability of composites. They also provide uniform properties in all directions, unlike fiber reinforcement. Flakes and spheres also provide high electrical and thermal resistance.

PROPORTION OF REINFORCEMENT TO MATRIX

The proportion of the reinforcement to matrix is critical to the strength of the composite. The relationship is a curved line. The strength and stiffness of the composite increase as more reinforcement is added, but only to a point. For example, in fiber-reinforced polymer matrix composites, the strength falls drastically if the volume of reinforcement exceeds about 60%. The composite loses strength when the volume of reinforcement is too high because the matrix cannot thoroughly wet all the reinforcement. This results in a weaker bond.

ORIENTATION OF REINFORCEMENT

Composites can be designed to have different properties in different directions. The strength of a composite is related to the orientation of the reinforcement. Aligning all fibers parallel to one another provides the highest strength when stressed parallel to the direction of the fibers. This unidirectional character is referred to as *anisotropy.* To obtain a degree of *isotropy,* where properties are equal strength in all directions, fibers must be arranged in multiple directions. However, this reduces the unidirectional strength of a composite.

For example, a mat made of unidirectional fiber composite may have a tensile strength of 300,000 psi. If two layers of the mat are placed in two directions, such as 0° and 90°, the tensile strength in either direction is about 1/3, or 100,000 psi. If three layers are placed at 0°, 45°, and 90°, the tensile strength is approximately one-sixth of the unidirectional fiber mat.

INTERPHASE ADHESION

The relationship of interphase adhesion to the properties of the composite is very important. The wetability of the reinforcement is critical in determining the mechanical properties of the composite. The more readily the matrix is able to wet the reinforcement, the better the bonding between reinforcement and matrix.

A strong interphase bond is especially important for fiber-reinforced polymer composites. However, a weaker bond is more desirable in ceramic matrix composites. The weaker bond causes the reinforcement to pull loose from the matrix when a shock causes the matrix to crack. This tends to stop the crack from propagating.

STRUCTURAL SHAPES

Metals, woods, and plastics are available in shapes and sizes standardized by professional organizations. Composites or "advanced materials," however, take many forms, depending on specific applications. This is one of the major advantages of composites over conventional materials. In the past, the unique properties of a material greatly influenced the design of the final product. Composites, on the other hand, allow the material to be engineered to meet the performance requirements of the design. The materials and the product are designed in an integrated manufacturing process.

FABRICATION

A wide variety of fabrication processes are used to transform primary composite materials into finished products. Many of the processes are adapted or the same as those used to process conventional materials. One of the earliest processes used and the least complicated is hand lay-up. Hand lay-up is commonly used for fiber-reinforced fabric, Figure 12-14. During this process fiber-reinforced fabric is placed in a mold. Resin is then poured, sprayed, or brushed into the fibers to provide stiffness.

Other processes include compression molding, injection molding, and cold stamping, Figure 12-15. Figure 12-16 shows a sheet molding process used to fabricate automotive body components.

TESTING AND ANALYSIS

Depending on the application, a composite product may require different tests to ensure it meets design requirements. These tests generally include structural analysis and testing of mechanical, physical, and chemical properties. Structural analysis is used to observe the matrix structure, reinforcement distribution, and interphase bond. These structures directly control the property of any composite.

Mechanical property tests normally include tension, compression, shear, bending and fracture toughness test. While performing these tests, special attention must be paid to the direction of reinforcement. Physical properties tested are electrical conductivity, thermal conductivity, and coefficient of thermal expansion. They are critical for certain applications which require high conductivity and low thermal expansion. Chemical properties basically deal with environmental stability or corrosion resistance. These criteria are often necessary in order to determine the feasibility of composite applications under critical conditions.

Figure 12-14.
Hand lay-up of a reinforced fabric composite involves manually applying a resin to a reinforcement.

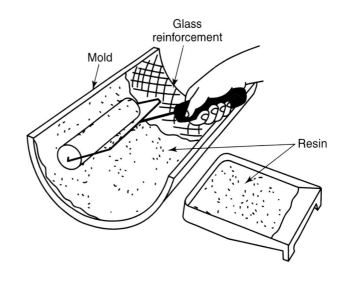

Figure 12-15.
Injection molding, compression molding, and cold stamping are used to form plastic composites.

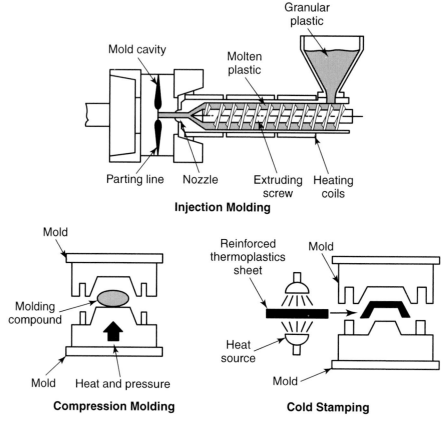

Figure 12-16.
The sheet molding process forms a plastic composite sheet.

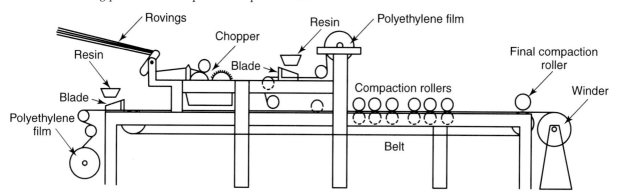

The field of composites is relatively new and much of it has been experimental in nature. Therefore, standards for this field are still being refined. However, when available, standards should be followed in performing composite testing and analysis. Several groups collect historical data on performance levels of well established composites. Committees have also been formed and the groups responsible for establishing standards are being identified. The American Society for Testing and Materials (ASTM) has established an advanced ceramics committee (C28). This committee is reviewing properties, performance, characterization, processing, terminology, design, and evaluation. ASTM's Committee on High Modulus Fibers and their Composites (D30), and Committee on Plastics (D20) are the primary sources for standardized test methods for PMCs. As composites becomes a more established field, more standards will be created and accepted.

Trade associations, such as the United States Advanced Ceramics Association (USACA) and the Suppliers of Advanced Composite Materials Association (SACMA), are also working with ASTM to establish standards. The Department of Defense (DOD) has initiated a program for standardization of composite technology. The DOD, through the US Army, is responsible for integrating diverse standards for composites by gathering standardized test methods into the *Military Handbook 17 (Mil 17)*. Separate test methods are developed when necessary. Other groups involved with establishing standards for composites include the Composites Manufacturing Association of the Society of Manufacturing Engineers (CMA SME), American Society for Metals (ASM), Society for the Advancement of Material and Process Engineering (SAMPE), and Society of Plastics Industry (SPI). Several international organizations are also pursuing advanced materials standards.

SUMMARY

The composites field is relatively new. The majority of composites and advanced materials have been used in high-performance military or aerospace systems, such as missiles, fighter aircraft, and space shuttles. However, manufacturers of a wide variety of commercial products have recognized the potential of these materials for existing and proposed products. Efforts to reduce the costs of the primary materials and improve the manufacturing processes for fabricating composite components are being made by many public and private organizations.

IMPORTANT TERMS

Anisotropy	Interphase
Aramid Fibers	Isotropy
Aspect Ratio	Matrix
Carbon/Graphite (CG) Fibers	Metal Matrices
Ceramic Matrix	Particulates
Composites	Polymer Matrices
Coupling Agents	Pyrolysis
E Glass	Rule of Mixtures
Fiber	S Glass
Fiberglass	Sphere
Flake	Wetability
Interface	Whiskers

QUESTIONS FOR REVIEW AND DISCUSSION

1. List factors that created a need for composites.
2. How have composites changed the design process?
3. Define composite.
4. List the three components of a composite material. Describe the unique contributions of each to the composite.
5. When were glass fibers introduced to commercial markets?
6. Describe the physical property differences between E glass and S glass.
7. What is the commercial name of the first aramid fiber material used in composites? What company developed it?
8. How are composites generally classified?
9. Describe the basic physical differences between thermosetting and thermoplastic resins.
10. What potential advantages do thermoplastics have over thermosetting resins?
11. How can the toughness of ceramic matrix composites be increased over conventional ceramics?
12. Define aspect ratio. Describe its effect on properties of composites.
13. How does the proportion of the constituents of a composite affect the strength of the composite?

FURTHER READINGS

1. ASM International. Engineering- Materials Handbook, vol. 1. Composites. Metals Park, OH 44073: ASM International (1987).
2. Agarwal, B. D. and Broutman, L. J. Analysis and Performance of Fiber ComRosites. New York: John Wiley and Sons (1990).
3. Schwartz, M. M. Composite Materials Handbook. New York: McGraw-Hill Inc. (1992).

INTERNET RESOURCES

www.sme.org
    Society of Manufacturing Engineers

www.sampe.org
    Society for the Advancement of Material and Process Engineering

www.composite.about.com
    General directory of web sites related to composites

www.asm-intl.org
    American Society for Metals

www.msel.nist.gov
    Material Science and Engineering Lab of the National Institute for Standards and Technology

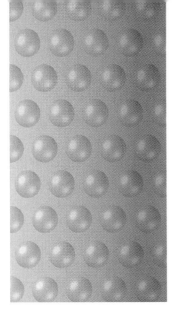

Chapter 13

Polymer Matrix Composites

KEY CONCEPTS

Upon completion of this chapter, you should understand:
➢ Thermoplastic matrices.
➢ Thermoset matrices.
➢ Common reinforcements used in polymeric composites.
➢ Manufacturing processes of polymeric composites.
➢ Properties and applications of polymeric composites.

Composites consist of matrix and reinforcement. The matrix is the "glue" which bonds fibers, whiskers, or particulate reinforcement together to form the composite. Polymers are the most widely used matrix materials for composites. Their major advantages are low cost, light weight, excellent processability, and chemical resistance. However, their applications are usually limited by their low strength, low modulus of elasticity, and low melting temperature. Most polymeric matrices also degrade by extended exposure to ultraviolet light and some solvents.

As discussed in Chapter 7 and Chapter 8, plastics can be classified as thermoplastics or thermoset according to their structure and behavior. Thermoplastic polymers consist of linear or branched-chain molecules. They soften or melt on heating. Thermosets have cross-linked or networked structures. They do not soften, but decompose, if heated to high temperature.

Thermosets are usually more rigid than thermoplastics and exhibit generally better high-temperature performance. However, some high-performance thermoplastics, such as polyetheretherketone (PEEK), are now equal to the most common thermoset in terms of temperature capability.

Thermosets require curing to create the final product. Thermoplastics, on the other hand, are heated, molded to desired shape, and then cooled. As a result, thermosets usually require much longer processing times than thermoplastics, resulting in high processing costs.

Since thermoplastics have a low processing cost, research tends to focus on these plastics. Although thermosets have been the principal matrix for composites, thermoplastics are increasingly used in many applications.

THERMOPLASTIC POLYMER MATRICES

Usually, a thermoplastic composite is simply an enhanced or filled polymer. Many thermoplastics used in composite materials can be divided into two categories. The first group consists of conventional thermoplastics. These have been used with short-fiber reinforcements for many years in industry. The second group consists of newer high-performance thermoplastics developed specifically for advanced composites.

CONVENTIONAL THERMOPLASTICS

Chapter 7 discussed several conventional industrial thermoplastics, including polyethylene, polystyrene, polypropylene, acrylics, and engineering resins. Refer to that chapter for explanations related to thermoplastics and their applications. Examples of these thermoplastics can be seen everyday. These plastics can be reinforced by fibers to improve their strength and rigidity. Reinforced plastics have been used for many years. Their use continues to increase at a phenomenal rate. The technology of using reinforced conventional thermoplastics is becoming more accessible and the performance of the resulting materials better.

Many common thermoplastic resins are available from manufacturers with reinforcements included in the resin. The most common fiber reinforcements in these resins are chopped glass fibers. The most common particulate reinforcements in these resins are minerals. In some cases, resin manufacturers denote the filled thermoplastics with a special name, such as DuPont's Minlon®. In other cases, they simply refer to the compound composition, such as 20% fiberglass-filled nylon. Common reinforced thermoplastics are nylon, acetal, polytetrafluoroethylene (PTFE), polysulfone, polyvinylchloride (PVC), acrylonitrile butadiene styrene (ABS plastics), polycarbonate, impact polystyrene, polyphenylene oxide, and polyethylene terephthalate (PETE). The resulting composites are used for gears, structures, and other industrial applications.

The thermoplastic used is often dictated by characteristics associated with the resin. These characteristics may include cost, environmental resistance, processability, coefficient of friction, resistance to creep, or many others. For example, conventional thermoplastics are often used for automotive body parts. In this application, a resin is chosen on the basis of structural integrity, ease of painting, ease of processing, cost, and environmental resistance. Fiber reinforcement is added to improve mechanical strength and rigidity of the resin.

HIGH-PERFORMANCE THERMOPLASTICS

Since the early 1980s, polymer chemists have developed advanced thermoplastic matrices suitable for high-performance composites. The result was a new class of composites with short processing times, superior damage tolerance, and adequate solvent resistance. *Damage tolerance* is the ability of a structure to function adequately after a flaw has been introduced.

Current high-performance thermoplastic matrices are characterized by a high glass transition temperature. This allows good mechanical performance at much higher temperatures than conventional thermoplastics. High-performance thermoplastic matrix composites have better toughness or impact resistance than thermoset polymers. This is very important in the aerospace industry. High-performance thermoplastic matrix composites generally have low manufacturing costs, high fracture toughness, good damage tolerance and impact resistance, good resistance to cracking, and easy quality control.

However, for practical applications of high-performance thermoplastic composites, some problems have to be overcome. These problems include difficulty in making and maintaining high quality of prepreg, cost of tooling at high temperature, and indeterminate fatigue and creep behavior. Prepreg is discussed in detail later in this chapter.

THERMOSET POLYMER MATRICES

Thermosets have traditionally been the principal matrix material. They still account for more than 80% of all reinforced plastics. In addition, thermosets are the matrix for essentially all commercial advanced composites. Thermoset matrices available include polyester, epoxy, polyimide, and phenolic polymers.

Molecule chains of thermoset polymers are interconnected with chemical bonds. This is known as cross-linking. Cross-linking is promoted by chemicals or by heat. In other words, curing is required to achieve the cross-linking and the final product.

POLYESTER RESINS

For many years, polyesters have dominated the market for commercial fiberglass-reinforced composites. These composites are used so much that the term "fiberglass" refers to the entire composite material. For example, a boat made from a fiberglass-reinforced polyester composite is simply called a "fiberglass boat." Polyesters are generally the lowest-cost matrices and can be used in a variety of processes to make composites. These factors led to their wide usage. However, compared to other available polymers, polyester has lower temperature capability, lower weather resistance, and lower physical properties. This has limited polyester use in advanced composites.

Cross-linking of polyester resins takes place by a polymerization reaction. This reaction requires a starting agent, such as styrene. The starting agent is also known as the *initiator.* Styrene reacts with polyester chains to produce cross-links. This entire cross-linking reaction is called *curing.* This forms a large network of interconnected polymer in which styrene serves as the cross-links (bridges) between molecules of the polymer.

Curing is only desired while the product is being formed. However, occasionally the resin mixture begins to cross-link spontaneously without initiators. Because of this, an inhibitor is added to the resin mixture. The *inhibitor* is a special substance that prevents spontaneous cross-linking.

Commercial polyesters are generally shipped and stored as a resin system. A *resin system* is the polyester resin, curing agents, fillers, and inhibitors all in one container. The initiator is stored in a different container. The contents of the two containers are mixed in appropriate proportion and the temperature raised when curing is desired.

Even with inhibitors, resins can eventually cross-link. Therefore, commercial polyester resins have a limited life span. The amount of time an unmixed resin can be stored before cross-linking occurs is called *shelf life* or *storage life*. The amount of time a resin can be stored after the initiator is added is called the *pot life*. Usually, a viscosity test is performed to see if the resin system is still acceptable for use. If the viscosity is too high, this suggests cross-linking has begun. In this case, the resin mixture must be discarded.

Occasionally, the curing rate is too slow. This may be the result of the presence of an inhibitor or simply because the chemicals need to react faster. In these instances, *accelerators* might be added to expedite the curing process. Typical accelerators are cobalt naphthenate, diethyl aniline, and dimethyl aniline.

EPOXY RESINS

For more demanding structural applications, epoxy resins are the preferred matrices. The quantity of epoxy-reinforced composites is small compared to polyester-reinforced composites. However, epoxy is used for nearly all demanding applications, such as aircraft or aerospace structures and printed circuit boards. Epoxies are used in these applications because of their excellent adhesion, high strength, low shrinkage, corrosion resistance, and processabilty.

Unlike many other polymers, the chain length of epoxy molecules before cross-linking can be as short as ten molecules. When cured, these molecules cross-link to form a three-dimensional network where the initiator becomes an integral part. The final properties of an epoxy-reinforced composite depends on the type of epoxy and the type of initiator. A wide range of properties can be obtained by varying the resin-to-initiator ratio and by using different initiators. This allows the composite performance to be tailored to specific application needs.

High-performance aircraft applications frequently use prepreg. *Prepregs* are fabrics pre-impregnated with resin. Prepregs can be produced using epoxies and curing systems that do not react at freezer temperatures, yet can be cured quickly at high temperatures. The low-temperature storage provides a long shelf life. These composites can provide service temperature capabilities of 250°F (120°C) to 350°F (180°C).

Epoxy systems have also been developed for rapid curing even under cold conditions. An example of this application is gluing reflectors to a road surface in winter. In general, temperature-related properties of a cured resin are dependant on the curing temperature. For example, resins cured at room temperature soften above room temperature. On the other hand, a resin that must withstand high service temperatures generally requires a high curing temperature.

A high degree of cross-linking is needed to achieve high temperature performance and corrosion resistance. This typically reduces the toughness of epoxy resin. However, recent advances have boosted the toughness of epoxy to equal or above other matrices, such as thermoplastics. This has been achieved with little loss in high temperature performance. In addition, these high-toughness epoxies can be processed like conventional epoxy resins. This means existing facilities and equipment can be used for the new resins.

POLYIMIDE

Polyimide resins are more expensive and less widely used than polyesters and epoxies. However, they are preferred when optimum thermal stability at high temperatures is required. Polyimide resins can be used at temperatures as high as 700°F (370°C). Epoxy resins, on the other hand, can generally be used up to 400°F (205°C). Polyimides are also inherently combustion-resistant. Currently, polyimide composites are used primarily in aerospace applications.

Polyimide can be formed by condensation polymerization. This type of polymerization is discussed in Chapter 8. The processing time is generally much longer for polyimide than other resin systems because of condensation polymerization.

PHENOLIC

Phenolic has been used for many years as a general, unreinforced thermoset plastic for electrical switches, junction boxes, molded automotive parts, consumer appliance parts, handles, and even billiard balls. However, phenolic resins have limited, but important, uses for composites. The principal applications for phenolic-resin composites are rocket engine nozzles and nose cones. Other uses include high-temperature aircraft ducts and muffler repair kits.

MANUFACTURING

Polymer matrix composites can be manufactured in many ways, including manual and fully-automated operations. Due to the ever increasing applications of polymer matrix composites, the manufacturing processes have experienced significant advances.

MANUAL LAY-UP AND PREPREG

The simplest technique is called lay-up or lay-up molding. This was probably the first technique used to make modern composites. Lay-up is used mostly for polyester and glass fiber composites. However, some epoxy-fiberglass composite parts are also produced in the same way.

In *lay-up,* fabric or mat is saturated with liquid resin. Layers are placed onto a shaped surface or mold by hand. Layers are added until the desired thickness is obtained. Pressure is applied to remove trapped air and provide uniformity. The pressure is normally applied by hand rolling, wiping with a squeegee, or using vacuum bagging. Vacuum bagging is discussed in the next section. The product is then allowed to cure. Limitations of manual lay-up include low production volume, inconsistency in product thickness, and inconsistency in the ratio of fiber to resin.

Lay-up can also be done using prepreg. Prepregs are fabrics which have resin preimpregnated before lay-up. They are used in applications where the part performance is critical. The difficulty of handling resin and fiber can be overcome and better quality control of final products can be achieved by using prepreg. Many resins are available in prepreg forms.

VACUUM BAGGING

Vacuum bagging is a process of applying a vacuum to the lay-up to help compress the layers. This is helpful for manual lay-up, but is required for prepreg lay-ups. The vacuum provides the advantages of pressing the composite layers together and simultaneously withdrawing excess "volatiles." The "volatiles" might be residual solvent, unreacted resin, or trapped air. A vacuum bag assembly is shown in Figure 13-1.

FILAMENT WINDING

In a *filament winding* process, a continuous tape of resin-impregnated fibers is wrapped over a form to create the part. Successive layers are added at the same or different winding angles until the required thickness is reached. Figure 13-2 illustrates a common filament winding system.

In one sense, filament winding can be thought of as a type of automated lay-up machine. However, filament winding normally does not use prepreg. Filament winding usually impregnates the fiber using a resin bath. Therefore, the process is called *wet filament winding* or *wet winding.*

AUTOCLAVE CURING AND BONDING

Autoclave curing and bonding is a process where a part is cured in an autoclave after lay-up or winding. The autoclave is heated and pressurized, usually at 50 psi (340 kPa) to 200 psi (1380 kPa). This process allows a complex curing reaction to occur. Additional pressure allows higher density and improved removal of volatiles from the resin.

Figure 13-1.
A typical vacuum bag assembly. (SME)

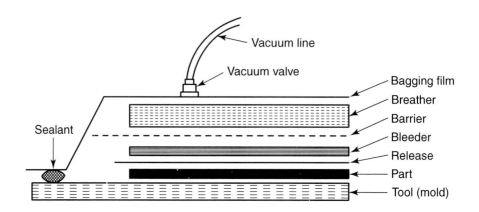

Figure 13-2.
A continuous tape of resin-impregnated fibers is wrapped over the mandrel to form a part. (SME)

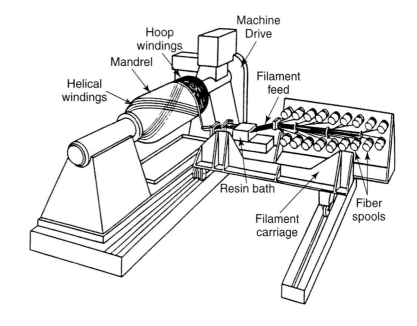

Figure 13-3 shows a typical autoclave system. Major elements of an autoclave system include: a vessel, a heated and pressurized gas stream, and a vacuum source. The vessel is where the part is placed and contains the pressure. The heated, pressurized gas is circulated uniformly within the vessel. An inert gas is typically used to reduce the danger of fire. The vacuum source compresses the layers of the composite. The autoclave also has a control system to control operating parameters. Many autoclaves have automated part loading/unloading systems as well.

Figure 13-3.
A typical large autoclave system. (Bondtech Treatment Technology)

The major drawback with an autoclave system is the high investment cost. Since it is a pressure vessel, it must conform to stringent pressure code regulations. Meeting these regulations often results in an expensive piece of equipment.

PULTRUSION

Pultrusion is a manufacturing process for producing continuous lengths of reinforced composite structural shapes. The basic concept of pultrusion is shown in Figure 13-4. However, pultrusion machines vary with part geometry.

Spindles, or "creels," hold the reinforcement for feeding into guides. A resin bath saturates, or "wets-out," the reinforcement with a solution containing the resin, fillers, pigment, catalyst, and other required additives. On exiting the resin bath, the composite is in a flat sheet form.

The composite is passed by a preform. A *preform* is an array of tooling that squeezes away excess resin and gently shapes the material before entering the die. In the die, the thermosetting reaction is activated by heat. The composite is allowed to harden and cure.

The part must be cooled before it is gripped by the "pull block" and removed from the die. This is to prevent cracking and deformation by the pull blocks. After exiting the die, the cured profile is cut to length.

COMPRESSION MOLDING

Compression molding is one of the oldest manufacturing techniques in the plastics industry. Traditionally, it has been used for molding rubber compounds and thermoset powders, such as phenolic and alkyd. With the recent development of high-strength sheet compounds and a greater emphasis on the mass production of composite materials, compression molding is receiving a great deal of attention.

Figure 13-4.
Pultrusion involves pulling reinforcement through a resin impregnator and a heated die to form a continuous length of reinforced composite. (Courtesy Morrison Molded Fiber Glass Company)

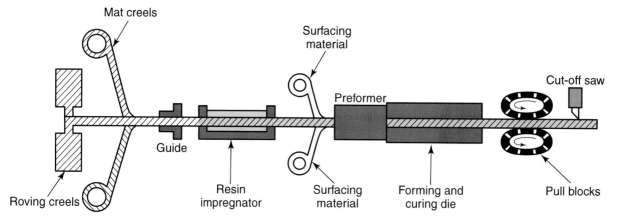

Compression molding has a number of advantages over the injection molding process. It is performed using relatively simple tools. The sprues, runners, or gates required for injection molding are not needed for compression molding. Consequently, very little material is wasted. Moreover, injection molding is limited to small quantities of fiber and fiber lengths of about 0.1″ (3 mm) or less. High quantities of fiber or long fibers can be easily handled in compression molding. Thus, better physical and mechanical properties can be achieved in compression molded parts. The pressure used in compression molding is lower than that in the injection molding. This means a lower capacity press is required, making compression molding better than injection molding for parts with large surface areas.

Figure 13-5 shows the schematic of a typical compression molding process, which can be divided into three basic steps. These steps are charge preparation and placement, mold closing, and curing.

Charge preparation and placement

In the charge preparation and placement step, a stack of sheet molding compound is placed in a preheated mold. Sheet molding compound is discussed later in this chapter. The stack of sheets is called the *charge.* The sheets are first die-cut into the desired shape and size. They are then stacked into a charge outside the mold. The charge weight is measured just before it is placed into the mold. The charge is then placed in the mold.

Ply dimensions are selected to cover 60% to 70% of the mold surface area with the charge. The charge location in the mold is a key factor in determining the quality of the molded part. The location influences fiber orientation, porosity, and other defects and, therefore, the final property of the composite.

Figure 13-5.
In a typical compression molding process, charges are cut and placed in a mold. Then, the mold is closed to form the part.

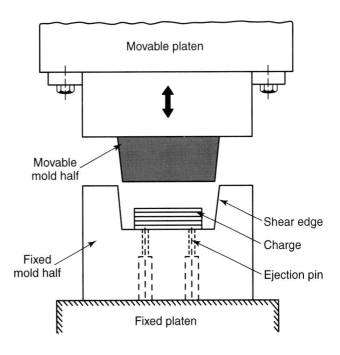

Mold closing

After the charge has been placed in the bottom mold, the top mold is moved down quickly until it contacts the top surface of the charge. Then, the top mold is closed at a slower rate. As temperature in the charge rises, the viscosity of the sheet molding compound is reduced. As the mold continues to close and increase pressure, the sheet molding compound flows to the cavity extremities. This forces air in the cavity to escape through the mold edges or vents.

The molding pressure based on the projected part area ranges from 145 psi to 5,800 psi (1 MPa to 40 MPa). This depends on the part complexity, flow length, and resin viscosity at the mold temperature. The common mold surface temperature is approximately 300°F (150°C). Both top and bottom molds are externally heated to maintain the mold surface temperature.

Curing

After the cavity has been filled, the mold remains closed for a predetermined period to assure a reasonable level of curing and ply consolidation through the part. The curing time may vary from one to several minutes. The amount of time depends on several factors, including the resin-catalyst-inhibitor reactivity, part thickness, and mold temperature.

At the end of the curing period, the top mold is opened and the part is removed from the bottom mold. The part is allowed to cool outside the mold. The mold is then prepared for the next part.

SHEET MOLDING COMPOUND

Sheet molding compound (SMC) is a continuous sheet of ready-to-mold composite material. It contains fibers and mineral fillers dispersed in a thermoset resin. The resin is in an uncured, but highly viscous state. SMC is cured and transformed into a finished product by compression molding.

The major components in SMC are resin, fibers, and fillers. Other components are used in small quantities, such as catalyst, thickener, inhibitor, mold release agent, and low profile additives. Figure 13-6 lists the composition of a typical sheet molding compound.

Figure 13-6.
This table shows the composition of a typical sheet molding compound.

TYPICAL COMPOSITION OF SHEET MOLDING COMPOUND (SMC)	
Material	**Amount (wt.%)**
Resin paste	
Unsaturated polyester	10.5
Styrene monomer	13.4
Low profile additive	3.45
Filler	40.7
Thickener (MgO)	0.7
Catalyst	0.25
Mold release agent	1
Inhibitor	Trace amount
Fiber	
E glass (25 mm long)	30

Principal steps in the production of SMC sheets include mixing, compounding, compaction, and maturation. Figure 13-7 shows the process of producing SMC. As the first step, a resin paste is prepared in the mixing step. Fibers are added to the resin paste layers in the compounding step. This is followed by impregnation (wetting) of fibers in the compaction step. For crossed (X) continuous fiber reinforced sheet, the compounding and compaction steps are carried out on a filament-winding machine. Finally, a maturation step is used to increase the resin paste viscosity in the SMC sheet. This is done to achieve a uniform and reproducible value suitable for the molding operation.

When it is ready for molding, the required length of SMC sheet is cut from the roll. The plastic film backing is removed before the SMC sheet is loaded into the compression molding press. Parts as large as car bodies can be made in this manner. SMC processing pressures may be as high as 5000 psi (34 MPa) with temperatures in the range of 250°F to 320°F (120°C to 160°C) and mold cycles of several minutes.

Sheet molding compounds can be used to produce many structural composite parts of complex shapes in relatively short molding time. Following are the principal advantages of an SMC composite.

- ◆ **Variety.** Fiber content and type in SMC can be easily controlled to produce a variety of mechanical and physical properties in the finished part.
- ◆ **Part consolidation.** SMC parts can be compression molded with ribs, bosses, curvatures, holes, and inserts incorporated. Using these design features, several components can be consolidated into one part and several secondary operations or assembly steps can be eliminated.
- ◆ **Light weight.** SMC composites have a lower specific gravity than structural metals. As a result, SMC parts are considerably lighter than metallic parts on an equal volume basis. The strength-to-weight ratio of SMC composites is comparable to that of many structural metals.

Figure 13-7.
Sheet molding compound (SMC) is produced by chopping glass fibers onto a sheet of plastic film, usually polyethylene, to which a resin paste mixture has been applied.

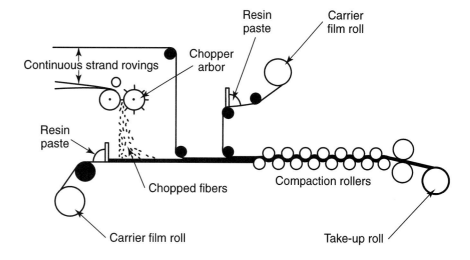

♦   **Dimensional stability.** Parts molded with SMC exhibit better dimensional stability than thermoplastic composites over a wide range of temperatures and other environmental conditions. The change in volume during curing can be controlled by the addition of various shrink-controlling additives. Furthermore, the coefficient of thermal expansion of SMC composites can be controlled to match that of steel or aluminum. This makes SMC parts compatible with steel or aluminum parts.

The transportation industry utilizes the largest volume of SMC composites. SMC composites are also used in appliances, such as washing machine doors and refrigerator housings. Furniture, such as chairs and tabletops, are often made from SMC composites. Business machines, such as computer housings, often use SMC composites. A variety of construction applications, such as door panels, use SMC composites.

INJECTION MOLDING

Injection molding is a process where a molten polymer is injected into a closed mold. The polymer solidifies in the mold, taking the shape of the mold cavity. The mold is then opened and the part is ejected and removed. Of all plastic processing methods, injection molding is second only to extrusion in application volume. Injection molding of plastics is discussed in Chapter 7. Refer to that chapter for details on injection molding of unfilled or unreinforced plastics.

Thermoplastics

Most thermoplastic resins are available in several filled and/or reinforced forms. Terms *filled* and *reinforced* indicate that a discontinuous and rigid phase has been blended into the polymer to make a polymeric matrix composite. When the aspect ratio is close to 1, the second phase is referred to as "filler" or "filled." If the aspect ratio is much larger than 1, as in the case with fibers, the term "reinforcement" or "reinforced" is used. Mechanical properties of thermoplastics are improved by reinforcement. For example, glass fibers provide a much higher rigidity than unfilled polypropylene at room and high temperatures. Flexural modulus of polypropylene homopolymer is increased from 246,500 psi (1.7 GPa) to 942,500 psi (6.5 GPa) when the composite is filled with 30% glass fibers. The softening temperature under load is increased from 149°F (65°C) to 298°F (148°C).

Reinforcing material can be either of fiber or flake. In practice, fiber reinforcements are almost exclusively used. Glass fibers dominate this market segment. Carbon or aramid fibers provide higher stiffness at lower weight, but their use in large volume applications is limited due to their high cost. Flake reinforcements, such as talc, mica, or glass flake, are also available. They can be used where stiffness and isotropy are required. Mica has found applications as a low-cost stiffener of certain commodity plastics where lower impact strength is acceptable.

Most thermoplastic composites are made in two steps. The first step is known as compounding. In this step, fibers or fillers are blended into the molten resin using a single-screw or twin-screw extruder. The extruded compound is then cut or made into pellets for further injection molding. The compound can be processed on an ordinary injection molding machine at a high production rate. However, this process has caused some problems for product properties and processing equipment. The fiber orientation in the molding is determined by the plastic flow, which may not be ideal. Also, the fiber can break during the compounding and injection processes. Finally, glass fibers are abrasive. This causes excessive wear of equipment and molds, especially at high temperatures.

Thermoset plastics

Injection molding of thermoset is similar in many aspects to the injection molding of thermoplastics. The molding compound is typically in the form of pellets or granules, usually gravity-fed from a hopper. The material is then moved forward by a rotating screw. The screw barrel is heated to approximately 200°F (95°C) at the front end. It is very important to avoid excessive temperatures in the barrel to minimize premature curing.

The heated viscous material is forced through the nozzle, sprue, and runner system into the mold cavity. The filling time may be as short as a couple of seconds for small parts to 10 seconds or longer for large parts. Once the mold cavity is completely filled, high pressure is maintained for several seconds to minimize any dimensional changes that might occur.

The mold is heated, which causes the material to cross-link and transform into a solid. The mold can be heated by circulating hot oil or steam. However, it is most commonly done using electrical cartridge heaters connected to temperature control units.

When the part is sufficiently rigid, the mold is opened and the molded part is ejected. The ejected part may continue to cure as a result of residual heat present in the part.

Various types of thermoset molding compounds are commercially available, including allyls, aminos, epoxies, phenolic, thermoset polyesters, thermoset polyimides, and silicones. Thermosets inherently possess excellent mechanical properties, heat resistance, and chemical resistance because of their three-dimensional network structure. Their ability to retain their performance characteristics and dimensional stability at elevated temperatures makes them particularly competitive and attractive over thermoplastic polymers for some applications.

RESIN TRANSFER MOLDING

Resin transfer molding (RTM) is developed from conventional reaction injection molding. In conventional reaction injection molding, no reinforcement is added. However, in resin transfer molding, a mold is loaded with reinforcement material before the resin is injected into the mold. A typical resin transfer molding process is shown in Figure 13-8. The mold is often subjected to vacuum to remove trapped air from the reinforcement. The reinforcement material is wetted by the pressure of the injection.

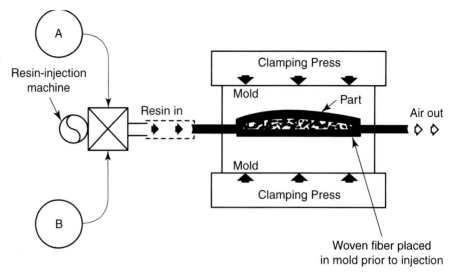

Figure 13-8.
In resin transfer molding, resins are injected slowly into a mold in which reinforcement has been loaded. (SME)

Reinforcements

Most standard reinforcement materials can be used for RTM. However, glass, carbon, and aramid fiber are the most commonly used. One important requirement of any reinforcement is that it retains its shape during subsequent injection. Therefore, reinforcements are generally stitched or bonded together and preforms are commonly used. Inserts of various types, such as screw receptacles and ribs, can be easily placed in the open mold while the fibrous reinforcement is placed in.

Resins

Most polymer resins can be used for resin transfer molding, including polyester and epoxy. Slow injection is essential for a good wetting of the reinforcement. The viscosity of the resin must be low enough for the fibers to be easily wetted. The resin should also have a pot life of about two hours so that injection can be slow without having the resin gel.

Mold

The design of the mold is the most critical factor in successful resin transfer molding. A mold must be constructed so the resin reaches all areas and resin concentrations are approximately the same throughout. This resin movement must be accomplished within the time allowed before curing. Moreover, the injection process should not move the reinforcement. The mold must maintain the shape of the reinforcement during the injection process.

RTM molds must be vented to allow air within the mold to escape. The vent should be small enough so resin will not flow out of the vent. A mold should also have a good temperature control so that resin curing can be optimized at the proper temperature.

PROPERTIES AND APPLICATIONS

The properties of any composite depend on the type of matrix and the reinforcement's concentration, form, distribution, and orientation. There are virtually unlimited combinations. It is impractical to list all the properties of different polymer composites. Therefore, this chapter looks at a few typical composites to illustrate the behavior of polymer matrix composite materials.

THERMOPLASTIC MATRIX COMPOSITES

Properties of thermoplastic composites differ significantly with fiber length. A general property comparison between short-fiber reinforced thermoplastic composite, long fiber reinforced composite, and unreinforced engineering thermoplastic is shown in Figure 13-9. As you can see, long fiber reinforced composites generally have the best properties. For example, the tensile strength of long fiber reinforced thermoplastic composite is approximately ten times higher than that of pure resin. However, short fiber reinforcement has only about twice the tensile strength. Figure 13-10 shows a typical application of thermoplastic composite materials.

The thermoplastic polyetheretherketone (PEEK) offers a superior combination of water resistance and toughness. This is very desirable in many applications. It has also shown good performance in both compression after impact and hot/wet compression.

Figure 13-9.
This chart shows various properties of thermoplastic composite materials. As you can see, long fiber reinforced composites have much higher values than others.

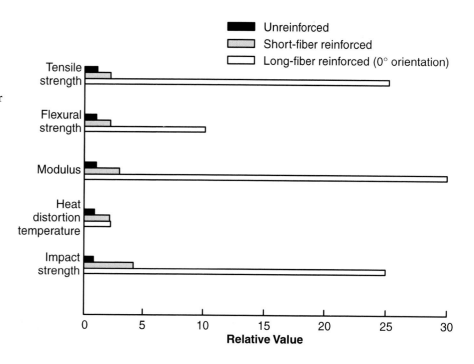

Polyamide (nylon) is another typical engineering thermoplastic used in composite form. Commercial resins have variations that may include alloys of polyolefin and other copolymers to tailor properties for specific uses. Short carbon or glass fiber reinforcements of up to 43% are used to improve properties in molding compounds. In general, this increases tensile strength by approximately 90% to 100%. For example, at room temperature the ultimate tensile strength of nylon is increased from 12,000 psi (83 MPa) without glass fiber reinforcement to 24,000 psi (166 MPa) when 33% short glass fibers are used. However, all nylons pick up moisture. Therefore, the final design properties and dimension change depend on the relative humidity. In general, high humidity tends to decrease the mechanical strength of nylon composites.

THERMOSET MATRIX COMPOSITES

Themoset matrix composites have dominated applications in industry due to their superior stability and mechanical properties. Various matrices can be selected for different service temperature applications. In general, higher service temperatures present more challenges to the composite material and, therefore, result in a higher-cost composite.

High-temperature thermoset matrix composites

Polyimide is typically used in a high-temperature thermoset matrix composite because of its characteristics at high temperature. The resin can be used at 600°F (315°C) for a long duration and at 900°F (480°C) for a short period. Polyimide can be reinforced with fillers and short fibers and used as a molding compound. In addition, continuous fibers or fabric can be pre-impregnated with polyimide resin and used for lay-ups, filament winding, and tape wrapping. At 482°F (250°C), the ultimate tensile strength of polyimide resin is 7,000 psi (48 MPa). A glass fabric polyimide composite can have an ultimate tensile strength up to 45,000 psi (310 MPa). The heat deflection temperature at 264 psi (1.82 MPa) is 582°F to 680°F (305°C to 360°C) for polyimide resin and 660°F (350°C) for 50% glass fiber reinforced composite. Polyimide thermoset composite also has excellent electrical properties. Major applications are in the aerospace and electrical industries.

High-strength, medium-temperature thermoset matrix composites

Epoxy resin systems are typically used where high strength is required at medium temperature. The resin can be used up to between 450°F to 500°F (230°C to 260°C) for short periods. Reinforced epoxy structures provide high strength-to-weight ratios and good thermal and electrical properties.

Filament winding and lay-up are used to produce advanced aircraft fuselages, wing and control surface panels, rocket motor cases, rocket nozzle structural shells, commercial pressure vessels, tanks, and pipes. Glass, carbon, graphite, quartz, and aramid fibers are used in molding compounds, hand lay-ups, and fiber/fabric prepreg composites for various applications.

The mechanical strength of epoxy composites decreases with increasing temperature. At room temperature, the tensile strength of epoxy resin with E glass reinforcement is in the range of 6,000 psi to 12,000 psi (41 MPa to 82 MPa) and the compressive strength is 17,000 psi to 50,000 psi (117 MPa to 345 MPa).

Medium-temperature thermoset matrix composites

Phenolic resin systems are typically used for medium-temperature thermoset matrix composites. Resins are usually impregnated in glass, carbon, and graphite cloth. They are used for tape wrapping or hand lay-up of aerospace components, rocket nozzle ablative, and insulation liners. Chopped fiber molding compounds are used mostly in automotive, appliance, and electrical component markets. High service temperatures, good electrical properties, excellent moldability, dimensional stability, and good moisture resistance make them suited for these applications. The in-service temperature of phenolic resins is 300°F to 450°F (150°C to 230°C), whereas that of E glass reinforced composites can reach 1000°F (540°C).

Phenolic resin can be compounded with chopped glass fiber or other fillers. It can also be incorporated in tapes for lay-up. Commercial forms are supplied as glass-fabric-phenolic tape, carbon-fabric-phenolic tape, and graphite-fabric-phenolic tape. The tensile strength of fiber-reinforced composite can be between 2,000 psi (14 MPa) and 10,000 psi (69 MPa) and compressive strength between 16,000 psi (110 MPa) and 38,000 psi (262 MPa).

Glass-cloth-fabric-phenolic thermoset composite has an ultimate tensile strength of up to 60,000 psi (420 MPa) and compressive strength up to 76,000 psi (524 MPa). The material strength decreases with increasing temperature.

Low-temperature thermoset matrix composites

Thermoset polyester resins are widely used for low-temperature applications. Unsaturated polyesters are the most extensively used type of thermoset resin because of the low cost, ease of processing, and good performance characteristics. They are generally combined with chopped, continuous, or woven glass fibers, as well as fillers and additives, to alter the properties to fit various applications.

The service temperature of polyester resin is in the range of 250°F to 300°F (120°C to 150°C). Adding 10% to 40% glass fiber increases the service temperature to 250°F to 400°F (120°C to 205°C). An unfilled cast polyester has a tensile strength 4,000 psi to 12,000 psi (27.5 MPa to 82 MPa). Polyester reinforced by glass fabric can have a tensile strength up to 85,000 psi (586 MPa). The compressive strength for unfilled polyester ranges from 12,000 psi (82 MPa) to 28,000 psi (193 MPa). The compressive strength of a polyester reinforced by glass fabric can be as high as 72,000 psi (496 MPa).

Thermoset polyester composites are widely used in transportation, construction, electrical, and consumer products. In addition, the versatility of polyesters allows them to be fabricated by a variety of processes. Through appropriate selection of the cross-linking initiator, these resins can be cured anywhere from room temperature to 350°F (175°C). Resin and glass fibers are combined at the mold in hand lay-up, spray-up, filament winding, pultrusion, and resin transfer molding. Figure 13-11 illustrates a typical application of glass-fiber-reinforced polyester composites in the construction of a structure.

Figure 13-11.
One of the four three-story turrets constructed of polyester fiberglass composite to house communications equipment on top of the Sun Bank building in Orlando, FL. (Photo courtesy of Strongwell Corporation, Bristol, VA)

SUMMARY

Polymer matrix composites are by far the most common composites used in industry. Polymer matrix materials are classified as thermoplastic or thermoset. Thermoset matrix composites dominate the composite market and applications. Most of the existing processing technologies for producing composites were developed for thermoset matrix composite. However, thermoplastic matrix composites have gained considerable attention recently due to the development of high performance thermoplastics and potential benefits offered by thermoplastics in processability.

Composite properties and, therefore, their applications are dependent upon the type, composition, size, distribution, and orientation of matrix and reinforcement. All these parameters are directly controlled by the manufacturing process. Thus, understanding the effect of process on the composite structure is essential to control the composite properties and to use these composites properly.

IMPORTANT TERMS

Accelerators	Preform
Autoclave Curing	Prepreg
and Bonding	Pultrusion
Charge	Reinforced
Curing	Resin System
Damage Tolerance	Resin Transfer Molding (RTM)
Filament Winding	Sheet Molding Compound (SMC)
Filled	Shelf Life
Inhibitor	Storage Life
Initiator	Vacuum Bagging
Lay-Up	Wet Filament Winding
Pot Life	Wet Winding

QUESTIONS FOR REVIEW AND DISCUSSION

1. What is a polymer matrix composite?
2. What are the two main advantages of using thermoplastics as a matrix for polymer composites?
3. What is a principal molecular difference between thermoplastic composites and thermoplastics?
4. What is damage tolerance?
5. What is initiator used for?
6. What is storage life?
7. What is pot life?
8. What is prepreg?
9. What is vacuum bagging for composite manufacturing?
10. What is autoclave molding?
11. What are the major components for a typical autoclave system?

12. Why is inert gas usually used in an autoclave system?

13. What is pultrusion? What is its main advantage and main application?

14. What is sheet molding compound (SMC)?

15. What are the main advantages of compression molding composites?

16. What type of fiber is suited for injection molding?

17. What is a compounding process for reinforced plastics?

18. What is resin transfer molding (RTM)?

19. What are the four types of thermoset matrix composites?

20. In general, what determines the properties of any composite?

FURTHER READINGS

1. Strong, A. B. <u>Fundamentals of Composite Manufacturing: Materials, Methods, and Applications</u>. Dearborn, MI: SME (1989).

2. Agarwal, B. D. and Broutman, L. J. <u>Analysis and Performance of Fiber Composites (2nd ed.)</u>. New York: John Wiley & Sons, Inc. (1990).

3. Carlsson, L. A. (ed.). <u>Thermoplastic Composite Materials</u>. Amsterdam: Elsevier (1991).

4. ASM International. <u>Engineering Materials Handbook, Vol.1: Composites</u>. Materials Park, OH (1987).

5. Mallick, P. K. and Newman, S. (eds.). <u>Composite Materials Technology: Processes and Properties</u>. Munich : Hanser (1990).

INTERNET RESOURCES

www.ornl.gov/orccmt/pages/homepg.html
Center for Composites Manufacturing Technology, Oak Ridge National Laboratory

www.advanced-composites.com
The Advanced Composites Group

www.thomasregister.com/olc/teel/pult_pro.htm
Teel Plastics: Pultrusion

www.strongwell.com/pult/pultrusion.htm
Strongwell

www.owens-corning.com/composites
Owens-Corning

oea.larc.nasa.gov/PAIS/ASM.html
NASA's Advanced Stitching Machine

www.ccm.udel.edu
University of Delaware, Center for Composite Materials

www.fibreglast.com
Fibre Glast: a supplier of glass fibers

www.iasco-tesco.com
IASSCO: a supplier of project materials

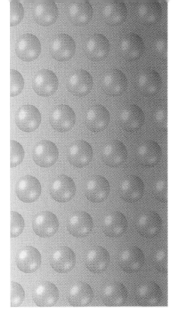

Chapter 14

Ceramic Matrix Composites

KEY CONCEPTS

Upon completion of this chapter, you should understand:
- ➢ The composition of ceramic matrix composites.
- ➢ The composition and function of the reinforcement materials in ceramic matrix composites.
- ➢ The composition and function of the matrix materials in ceramic matrix composites.
- ➢ The processes used to manufacture ceramic matrix composites.
- ➢ The properties of ceramic matrix composites.
- ➢ The applications of ceramic matrix composites.

Ceramic matrix composites (CMC) are generally made of a glass or ceramic matrix reinforced with fibers, whiskers, or particulates. They are sometimes defined as:
- ◆ randomly oriented ceramic whiskers within a ceramic matrix.
- ◆ continuous fibers oriented within a ceramic matrix.
- ◆ dissimilar particles dispersed in a matrix with a controlled microstructure.

Ceramic matrix composites are able to resist temperatures of 3000°F (1650°C) and higher. They are completely inorganic and highly inert. The thermal stability is better than all refractory metals. Refractory metals are metals whose melting temperatures are above 3000°F (1650°C), including titanium, tungsten, and tantalum. Other potential benefits of CMCs include higher hardness, increased thermal shock resistance, and significantly improved fracture toughness. A variety of ceramic particulates, whiskers, and fibers may be added to matrix host materials to form fracture- toughness, improved composites.

The primary focus of CMC research and development is to further increase the fracture toughness while substantially retaining the traditional properties of conventional ceramics. Many private and public research facilities, including the United States Department of Energy, the Oak Ridge National Laboratory, and the National Institute for Standards and Technology, are conducting extensive research to improve the properties of ceramic composites while reducing manufacturing costs. High production costs and lack of material reliability have been the greatest stumbling blocks to increasing the applications of ceramic matrix composites.

CMCs provide the most viable solution to the brittleness problems found in conventional ceramic bodies. Some success has also been achieved at improving fracture toughness without substantially affecting other desirable properties. This research and development is still being conducted.

CERAMIC MATRIX COMPOSITE CONSTITUENTS

By definition, composites consist of two or more materials or material phases combined to produce a material with superior properties to those of its individual constituents. The constituents are the matrix and the reinforcement. The interphase is also an important part of any composite, Figure 14-1. The matrix and reinforcement must be present in reasonable proportions to produce physical properties different than the individual material phases. In other words, if there is not enough reinforcement, the composite will have all the same properties of the matrix.

MATRIX

Using the literal definition of a composite material, Portland cement is by far the most popular matrix for ceramic matrix composites. The matrix is that material phase of a composite which glues or binds the reinforcement together. Portland cement is the matrix that holds reinforcing materials (sand and stone) together to form the CMC concrete. By volume concrete is the most used material of our manufactured and constructed world. This chapter, however, focuses on advanced ceramic composites. Therefore, Portland cement and concrete are not considered.

Figure 14-1.
The phases of a ceramic matrix composite (CMC) include the matrix, reinforcement, and interphase.

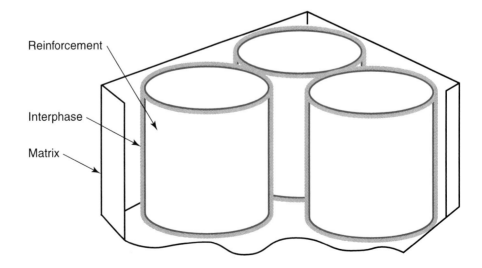

Reinforcement

Interphase

Matrix

Matrix material selection for ceramic composites is usually based on its thermal compatibility with the reinforcement. This compatibility, based on the coefficient of thermal expansion (CTE), is more critical in ceramic composites than it is in polymer matrix and metal matrix composites. If the CTE is substantially different between the matrix and the reinforcement, the composite will fail. The failure can be in a number of ways, such as the matrix separating from the reinforcement.

The most popular matrices used for CMCs are listed below.

♦   Lithium-Alumina-Silicate, or LAS (Li_2O-Al_2O_3-SiO)
♦   BMAS (BaO-MgO-Al_2O_3-SiO)
♦   Magnesia-Alumina-Silicate (Mg-Al_2O_3-SiO)

Of these matrices, lithium-alumina-silicate (LAS) is the most widely researched glass ceramic. LAS has a relatively low coefficient of thermal expansion (1.5×10^{-6}/k), which is a primary factor for its popularity as a ceramic matrix. Figure 14-2 provides general physical properties of selected ceramic composite matrices.

Figure 14-2.
This table shows various properties of ceramic composite matrices.

PROPERTIES OF CERAMIC COMPOSITE MATRICES				
Matrix Material	Melt Temperature	Thermal Expansion (10^{-6}/°C)	Fracture Toughness (Mpa $m^{1/2}$)	Young's Modulus (GPa)
Al_2O_3	2050°C	8.64	3.52	345
LAS	—	5.76	2.42	117
Mullite	1850°C	5.76	2.20	145
Pyrex	1252°C	3.24	0.08	48
TiO_2	1849°C	9.36	2.53	283
SiO_2	1610°C	0.54	0.77	76
Sic Sn	1980°C	4.32	4.94	331
SiC Hp	1980°C	4.32	4.94	414
ZrO_2 PS	2760°C	7.92	8.46	207
ZrO_2 FS	—	13.5	2.75	207
Si_3N_4 SN	1870°C	3.06	—	310
B_4C	2350°C	3.06	—	290
TiC	3140°C	8.46	—	427
Wc	2870°C	4.5	—	669

REINFORCEMENT

The reinforcement materials for ceramic matrix composites are classified by shape and size. Ceramic matrix composites reinforcement materials can be classified as fibers, whiskers, or particulates. The reinforcement itself is often classified as a ceramic; often glass fibers.

Fibers

The use of fibers in CMCs achieves increased toughness and more reliable composite ceramics. However, the improvement in fracture toughness also means reduced strength. The first engineered ceramic fibers were glass threads used to reinforce synthetic resins and polymers, Figure 14-3. This began in the 1930s. After WWII, silica ceramic fibers were developed as insulating materials.

In today's CMCs, the fibers are primarily compositions of oxides, nitrides, and carbides. Some of the most popular ceramic fibers used in CMCs are:

- Silicon Carbide SiC
- Silicon Nitride Si_3N_4
- Zirconium Oxide ZrO_2
- Aluminum Oxide Al_2O_3

In fiber form, these compositions have mechanical properties 2 to 3 times greater than those of polycrystalline monolithic structures of the same substances.

Carbon matrix composites have the highest potential use temperature of any CMC, greater than 3500°F (1930°C). The carbon matrix ceramic composites, however, require protective ceramic coatings as they oxidize readily in air at temperatures above 1100°F (595°C).

Ceramic oxide fibers can achieve tensile strengths as high as 2100 MPa and tensile modulus of 373 GPa. Ceramic oxide fibers have low thermal conductivity and high thermal stability.

Figure 14-3.
The first engineered ceramic fibers were glass threads used to reinforce synthetic resins and polymers.

Temperature limits can cause problems for ceramic fiber reinforcements. The high firing temperatures needed to consolidate the fiber/powder blends result in a loss of fiber strength. This effect is usually detrimental to applications requiring high reliability and mechanical properties.

Whiskers

Ceramic whisker reinforcements are typically high-strength single crystals with an aspect ratio of 10. Whiskers have much smaller physical dimensions than fibers. Whiskers are approximately 5 to 10 microns in diameter and less than 100 microns in length. Whisker strength increases as diameter decreases because the smaller diameter allows better whisker formation. Whiskers with a nominal diameter of .05 to 1.0 microns have strengths between 1 million to 2 million psi (6895 MPa to 13,790 MPa).

Whiskers are usually single crystals with mechanical properties 5 to 10 times greater than fibers of the same material. Improved mechanical properties, especially strength, are due to greater binding forces in adjacent atoms and fewer grain boundaries, voids, impurities, and imperfections.

Whiskers also display some elasticity and have withstood up to 3% strain without permanent deformation. This is an indicator of good toughness. Whiskers also tend to have fewer defects than fibers.

Silicon carbide (SiC) is the most common whisker reinforcing material for CMCs. CMCs are fabricated by uniaxial hot pressing. This process substantially limits size and shape capabilities. It also requires expensive diamond grinding to produce the final part.

Particulates

A reinforcement material whose dimensions are roughly equal is classified as a particulate. Therefore, particulate-reinforced CMCs are those reinforced with spheres, flakes, and rods, Figure 14-4. Concrete can be used as an example of particulate-reinforced CMC. As illustrated in Figure 14-5, the reinforcement is stone which has nearly equal dimensions in all axes. The stone is considered a sphere reinforcement.

Figure 14-4.
Reinforcement can be in the form of spheres, rods, or flakes.

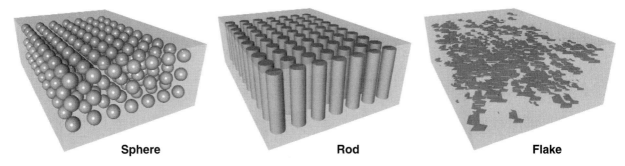

Sphere   **Rod**   **Flake**

Figure 14-5.
Concrete is a particulate-
reinforced CMC.

Flake-reinforced composites make excellent insulators. Composites reinforced with sphere-shaped particulates resist failure due to sudden shock loading. They also exhibit excellent compressive strength. In addition, they are dimensionally stable and exhibit a low coefficient of thermal expansion.

Particulate-reinforced ceramics are excellent engineering materials for cutting tools, seals, bearings, valves, heat engines, and medical implants. A wide range of additional applications are currently being developed as well.

MANUFACTURING PROCESSES

The development of CMC has been much slower than polymer and metal matrix composites. One reason for this is the forming or manufacturing processes have been more difficult to develop because of the high temperatures needed. Other physical properties common to all ceramics, such as brittleness, make processing and post processing difficult as well. The following is a list of forming and post-forming processes used to manufacture ceramic matrix composites.

- Laser Beam Machining
- Electrical Discharge Machining (EDM)
- Diamond Grinding
- Hot Pressing
- Powder Processing
- Chemical Vapor Infiltration (CVI)

LASER BEAM MACHINING

A laser converts electrical energy into highly coherent light, Figure 14-6. The unique properties distinguishing laser light from natural light is laser light is a single wavelength and the "waves" are perfectly parallel. The primary advantage of using a laser to shape CMCs is that it can cut the hard, brittle materials cost effectively.

Figure 14-6.
Laser beam machining uses a focused laser to machine parts.

Laser beam machining uses the focused light energy to separate or shape ceramic parts by vaporization. Both gas and solid state types of lasers are used for this process. With the addition of a computer numeric control system, lasers can cut complex shapes. However, the surface finish resulting from the laser beam machining process may require post processing to achieve an acceptable finish.

ELECTRICAL DISCHARGE MACHINING (EDM)

Electrical discharge machining (EDM) was developed primarily for the tool and die industry, Figure 14-7. EDM was originally developed for removing metal through controlled arcing between an electrode and the metal part to form the desired finished product. EDM has also been adopted for use in forming CMCs.

EDM removes material by storing electric current in capacitors and then discharging the current across a small gap between the electrode (cathode) and the work piece (anode). The arcing is controlled and takes place while the part is submerged in a dielectric fluid. The dielectric fluid helps flush out the chips from the cutting area and control the sparking. The addition of computer numerical control systems makes EDM even more versatile.

EDM can be used to shape any material that conducts electricity. CMCs that can conduct electricity are now being used for wear parts, extrusion dies, and cutting tools. EDM is capable of producing these parts in an exact shape. Complex shapes or special features can also be produced with EDM.

EDM produces surface finishes comparable to grinding. The cutting speeds can exceed conventional machining speeds of carbide materials.

Figure 14-7.
A—This diagram shows the EDM process. B—An arc, or spark, from the electrode causes the work to erode, thus forming the part.

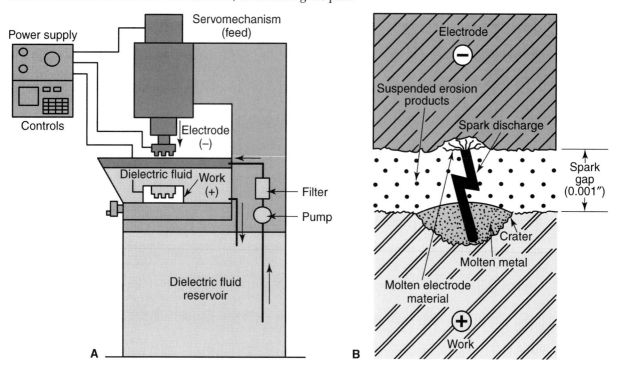

DIAMOND GRINDING

The demand for ceramic matrix composites with tighter tolerances and improved finishes is growing. However, customers are generally not willing to sacrifice productivity for the improved attributes. These factors have placed increased pressure and importance on diamond tools. These tools are capable of achieving high rates of stock removal while producing a finish that meets customer specifications. *Diamond grinding* is a common operation used, where a spinning abrasive wheel is moved across the part.

Surface quality is a combination of the width, depth, and spacing of machining grooves in the product and is measured in microns, Figure 14-8. Improved surface finishes are being achieved by increasing the rotating speed of the grinding wheel and by bonding the diamonds with a resin that easily crumbles. This type of resin is called *friable.*

Traditional bonding methods have been limited in controlling the relative height of the diamond particles over the grinding surface. The friability of the diamond bond causes the higher points of the grinding material to wear uniformly, minimizing the depth of the grooves during the grinding process. The increased friability of the diamond grinding material also improves the sharpness of the diamond particles. A diamond grinding material bonded with a highly friable resin causes new sharp edges to be exposed continuously. This aids in achieving more efficient material removal.

Figure 14-8.
A—The width, depth, and spacing of surface quality. B—How surface quality is specified on a drawing, in microns.

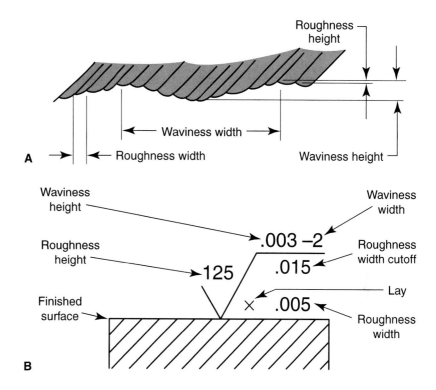

POWDER PROCESSING

Powder processing involves mixing of fine ceramic powders and applying force to create a composite. The powders are placed into a die and great force is applied. The powder takes the shape of the part (die). The pressed forms are then heated, causing the powders to fuse.

The first step in powder processing is matching a matrix to a reinforcement material. Incompatibilities between the constituents, such as different melting temperature, chemical reactions, or coefficients of thermal expansion, may cause early failure of the CMC component. Selecting the matrix and reinforcement requires technical knowledge of the properties of the powders. Size distribution, shape, specific area, density, electrical charge, and other factors must be considered in making the correct decisions.

A primary objective in correctly selecting constituents and properly mixing them is to create a composite with a minimum of voids and not very porous. Properly selecting and mixing CMC constituents is referred to as *packing.* Packing for optimum effect occurs when particle size distribution contains 30% volume of very small particles and 70% volume of large particles.

The powders are made using a variety of cold forming operations including cold isostatic pressing, tape casting, extrusion, compression molding, and injection molding. Once the powders are formed, a final consolidation and densification takes place at higher temperatures. The processes used for this purpose include sintering, hot unidirectional pressing and hot isostatic pressing.

The primary objective when forming a CMC is to achieve "near-net-shape" components. All ceramics, conventional and composites, are extremely brittle and difficult to machine. The closer to finish state the forming process can achieve, the less post-process shaping is required. Therefore, less time and resources are required.

HOT PRESSING

Hot pressing is a type of molding process that combines forming and firing. During this process, ceramic powders are heated to sintering temperatures and pressed together in a die, Figure 14-9. One advantage of this process is that it creates a very dense product with superior strength. The greatest disadvantage to this process is that it is slow.

A relatively recent high-temperature forming method used for forming advanced ceramics is *hot isostatic pressing (HIP)*. This process incorporates a high-pressure inert gas and high temperature to densify and form ceramic parts. During the process, ceramic powders or preforms are placed in a pressure vessel and an inert gas, such as helium, is injected into the chamber at the appropriate forming pressure. The furnace temperature is then raised to the desired bonding temperature and maintained for complete densification and bonding.

Figure 14-9.
This diagram shows a typical setup for hot pressing of ceramic matrix composites.

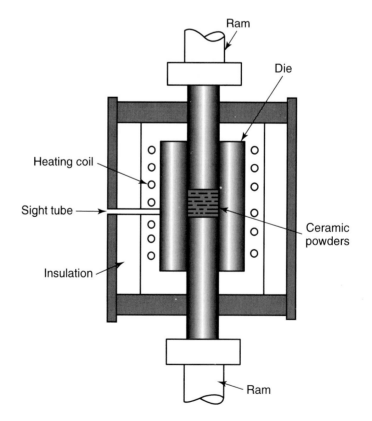

CHEMICAL VAPOR INFILTRATION

Chemical vapor infiltration (CVI) has become a popular way of agglomerating ceramic matrix composites. CVI involves introducing a gas into the composite at the reinforcing stage. The gas then decomposes to form the solid matrix. Carbides, nitrides, oxides, and borides have been used with silicon-carbide-based fibers and whiskers. Silicon carbide is commonly used as a matrix for CVI. The greatest disadvantage of CVI as a forming process for CMCs is the time and cost needed to complete a component.

PROPERTIES OF CMC

Generally, ceramics have superior strength at high temperature and low thermal conductivity. However, they are the hardest and most brittle of all manufacturing materials. A primary objective in developing ceramic matrix composites is to increase their toughness beyond conventional ceramics. Reinforcing ceramic matrices with fibers, particles, and whiskers improves their strength, toughness, and ductility.

By reducing the strength of the bond between the matrix and the reinforcement material, or interphase, the fracture toughness of CMCs can be increased. Long fibers in the ceramic or glass matrix can increase the fracture toughness by a factor of two over conventional ceramics. This is because the long fibers stop crack propagation. The toughest CMCs are continuous-fiber-reinforced glass matrix composites.

Reinforcements resist crack propagation to varying degrees in the following ways. First, when the progressing crack tip encounters a fiber, whisker, or particle, it is inhibited from continuing in that direction. Second, if the bond between the reinforcement and the matrix is not too strong, the energy the crack has to continue can be absorbed by pulling the reinforcement from its original location. Third, fibers can bridge a crack, thereby holding the two faces of the matrix together and preventing further growth of cracks.

There are properties, however, that may not be desirable. The reinforcements for CMCs can break down at intermediate temperatures. Additionally, if there is differences in coefficients of thermal expansion between the matrix and the reinforcement, thermal stress can occur during the cooling phase. This often causes cracking of the matrix.

SUMMARY

Ceramic matrix composites (CMC) generally consist of a glass or ceramic matrix reinforced with fibers, whiskers, or particulates. CMCs are able to resist temperatures of 3000°F (1650°C) or higher and exhibit improved fracture toughness when compared to conventional ceramic bodies. CMCs are inorganic and highly inert. Their thermal stability is better than all refractory metals.

The advantages of CMCs over other ceramics include high hardness, increased thermal shock resistance, and improved fracture toughness. High production costs and lack of material reliability have been the greatest roadblocks to increasing the applications of ceramic matrix composites.

The selection of matrix and reinforcement in CMCs is generally more critical than with other composites. A critical factor in matching the matrix and the reinforcement is thermal compatibility. This compatibility, based on the coefficient of thermal expansion, is absolutely essential in manufacturing a reliable composite. Poor thermal compatibility usually results in early component failure.

Several manufacturing processes have been used successfully to form ceramic matrix composite parts. Some processes have been used for years to form parts from conventional materials, e.g., injection molding, extrusion, jiggering. Others are relatively new or have been adapted from other areas. These include laser beam machining, electrical discharge machining (EDM), powder processing, hot pressing, chemical vapor infiltration (CVI), and diamond grinding. The primary objectives of these processes are to produce a near-net-shape part and do it cost effectively. Because CMCs are so hard and brittle, post-processing can be costly, time-consuming, and difficult.

Generally, CMCs have superior strength at high temperature, low thermal conductivity, and they are among the hardest materials. Although their development has lagged behind that of the polymer and metal matrix composites, major public and private national research laboratories are making significant progress in developing new advanced ceramic composites.

IMPORTANT TERMS

Ceramic Matrix Composites (CMC)

Chemical Vapor Infiltration (CVI)

Diamond Grinding

Electrical Discharge Machining (EDM)

Friable

Hot Isostatic Pressing (HIP)

Hot Pressing

Laser Beam Machining

Packing

Powder Processing

QUESTIONS FOR REVIEW AND DISCUSSION

1. Provide one of the three general definitions of ceramic matrix composites.
2. What are the advantages of ceramic matrix composites over conventional ceramics?
3. What three categories are used to classify reinforcement types for ceramic matrix composites?
4. What have been the greatest stumbling blocks to increasing the applications of ceramic matrix composites?
5. What is the thermal compatibility of matrix and reinforcement in ceramic matrix composites based on?
6. Why is thermal compatibility so critical for CMCs?
7. What are the advantages and disadvantages of the following reinforcement types? Fibers, Whiskers, Particulates
8. List four typical applications of ceramic matrix composites reinforced with particulates.
9. What are the primary objectives when forming ceramic matrix composites?
10. Describe the following manufacturing processes used in forming ceramic matrix composites. Laser beam machining, electrical discharge machining (EDM), powder processing, hot pressing, and chemical vapor infiltration (CVI).
11. What techniques are used improve the fracture toughness of ceramic matrix composites?

FURTHER READING

1. Bauccio, Michael. <u>Engineering Materials Handbook</u>. Materials Park, OH:ASM International (1994).

2. The Materials Information Society. <u>Engineered Materials Handbook: Desk Edition</u>. Materials Park, OH: ASM International (1995).

3. Lee, Stuart (ed). <u>International Encyclopedia of Composites</u>. New York:VCH Publishers (1990).

INTERNET RESOURCES

www.nist.gov
   National Institute for Standards and Technology

www.anl.gov
   Argonne National Laboratories

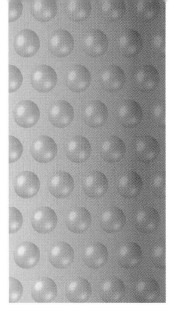

Chapter 15

Metal Matrix Composites

KEY CONCEPTS

Upon completion of this chapter, you should understand:
➢ The definition of metal matrix composites.
➢ Types of reinforcement used in metal matrix composites.
➢ Manufacturing processes for metal matrix composites.
➢ Properties of metal matrix composites.
➢ Advantages and applications of metal matrix composites.

Advancements in mechanical systems require performance characteristics beyond traditional materials. Conventional metals, alloys, and ceramics simply cannot meet the requirements. This fact has spurred the development of metal matrix composites (MMC). MMCs have performance characteristics beyond traditional materials.

A *metal matrix composite (MMC)* consists of a metal base reinforced with one or more other materials, such as graphite, ceramic, or metal. Examples of metal matrix composites stretch back to ancient civilizations. A copper awl from Cayonu, Turkey, dates to about 7000 BC. It was made by a repeated lamination-hammering process. This process resulted in high levels of elongated, nonmetallic inclusions that strengthened the metal. Among the first composite materials to see wide commercial and industrial use is the consolidated mixture of aluminum/alumina powders, starting in the 1920s. Interests in fiber-reinforced metal matrix composites mushroomed in the 1960s. These efforts were directed mainly at aluminum and copper matrix systems reinforced with tungsten or boron fibers. The first modern production of metal matrix composites was boron-aluminum tubes used as components on the space shuttle in the 1970s. MMCs rapidly developed during the 1980s, with a focus on aluminum-based composites reinforced with silicon carbide (SiC) or aluminum oxide (Al$_2$O$_3$) particles and short fibers.

To date, most MMCs are relatively expensive. In some instances, the MMC is also difficult to fabricate with high degree of reliability. However, with continued increase in development and use, the cost of metal matrix composites will decrease. Applications for MMCs continue to increase, especially under critical working conditions. This chapter discusses the fundamentals regarding matrices, reinforcements, manufacturing processes, properties, and applications of metal matrix composites.

MATRIX

Metal matrix composites were first developed to improve the performance of existing metallic alloys. Therefore, many of the matrix materials used today are alloyed metals. Common matrices for MMCs include aluminum alloys, titanium alloys, copper, and silver.

ALUMINUM

Aluminum is a good conductor of electricity and heat. It is also highly reflective to both heat and light. Aluminum alloys are highly corrosion resistant under many working conditions and are nontoxic. Certain aluminum alloys can be stronger than some structural steels.

The main advantage of aluminum alloy as a matrix for composites is light weight. Aluminum metal is widely used in the aerospace industry. There has been a significant increase in the use of aluminum alloy in automobiles. However, aluminum is not as strong as steel. An ideal solution is to design an MMC with an aluminum alloy as the matrix that has the strength and toughness of steels. To achieve this objective, industry has devoted tremendous efforts to develop aluminum matrix composites. Applications of aluminum matrix composites have been explored for structural components, engine blocks, and pistons.

TITANIUM

Titanium alloys typically have a high strength-to-weight ratio, making them attractive candidates for metal matrix composites. Titanium is about 57% of the weight of steel. Titanium has a density of 0.16 lb/in^3 (4.5 g/cm^3). Steel has a density of 0.28 lb/in^3 (7.86 g/cm^3). In addition, titanium alloys can have better mechanical properties than many alloy steels. Titanium also has higher stiffness than other light metals, such as magnesium and aluminum. Moreover, titanium alloys have excellent corrosion resistance in oxidizing acids, chloride media, and most neutral environments. Titanium retains adequate mechanical properties at temperatures up to about 1020°F (550°C).

Reinforcing titanium alloys with boron-based or silicon-carbide-based fibers increases stiffness and reduces weight. However, the composites may have lower fatigue strength due to many factors. These factors include residual stress and reaction between the fibers and the titanium matrix during high-temperature fabrication. Since titanium is very chemically reactive, certain protection of the fiber is needed during processing. Recent developments have significantly reduced the problem of low fatigue strength.

COPPER

Copper is widely used, both as pure copper and alloyed with other elements. Copper has high electrical and thermal conductivity, is easy to fabricate, and is low in cost. Conductivity of copper is adversely affected by almost all alloying elements. Therefore, alloying elements used to improve mechanical properties of copper alloys decrease the conductivity.

Copper matrix composites can be used for applications requiring high conductivity and high mechanical strength. For instance, copper-tungsten alloys containing carbon, Al_2O_3, SiC, and B_4C fibers have been produced. Recently, copper matrix composites have also been developed with refractory metal filaments as reinforcement. An excellent combination of high electrical conductivity and mechanical strength has been achieved with those composites. This is because copper and refractory metals, such as niobium, have very low solid solubility at room temperature. This means that an almost pure copper matrix can be formed with refractory metal filaments.

SILVER

Silver, in pure or alloyed form, is the most widely used material for a variety of electrical contacts, such as circuit breakers, connectors, and switches. Silver has the highest electrical and thermal conductivity of all metals at room temperature. Therefore, it can carry a higher electrical current without excessive heat. Annealed pure copper has a conductivity of 100% of the International Annealed Copper Standard (IACS). Pure silver has a conductivity of 104% IACS. Silver also has better oxidation resistance than copper, which makes silver more suitable for contact applications.

Silver is very ductile, which can be easily fabricated in various ways. However, this ductility is a drawback under certain conditions. In order to overcome this problem, silver-based composites have been developed with reinforcements such as cadmium oxide (CdO), tungsten (W), tungsten carbide (WC), nickel (Ni), molybdenum (Mo), and graphite.

REINFORCEMENTS

Reinforcements add strength to composite matrices. They are available in a variety of forms. Reinforcements can be in the form of continuous fibers, short fibers, and particulates.

CONTINUOUS FIBERS

Continuous fibers provide the most effective means of strengthening composite materials. They can be made of boron, SiC, carbon, or aluminum oxide.

Boron fiber

Boron filament is unique among composite reinforcement fibers available in production scale. It combines superior tensile strength, compressive strength, flexural strengths, high modulus of elasticity, and low density in one fiber. Figure 15-1 compares the mechanical and physical properties of boron fiber with those of 300 M alloy steel and high strength aluminum 2024 alloy. This figure shows that boron fiber has higher strength and modulus of elasticity (rigidity) than aluminum and steel alloys. Moreover, the density of boron fiber is even lower than that of aluminum alloys. Boron fiber is the first high strength, high modulus of elasticity reinforcing fiber used in metal matrix composite applications. It has been used in both aluminum and titanium matrices.

Figure 15-1.
This table shows the properties of boron fiber with alloys.

COMPARISON OF BORON FIBER PROPERTIES WITH ALLOYS			
Property\Material	Boron Fiber	Aluminum 2024	300 M Alloy Steel
Tensile Strength	500 ksi (3.45 GPa)	64 ksi (0.44 GPa)	290 ksi (2 GPa)
Modulus of Elasticity	58,000 ksi (400 GPa)	4,000 ksi (27.6 GPa)	30,000 ksi (206.9 GPa)
Density	0.090 lb/in^3 (2.5 g/cm^3)	0.098 lb/in^3 (2.7 g/cm^3)	0.283 lb/in^3 (7.84 g/cm^3)

Production of boron-reinforced aluminum matrix composites has been moderately successful. In the production of boron-reinforced titanium matrix composites, the boron fiber is exposed to severe processing environments that degrade the fiber's strength and stiffness. Surface coatings or diffusion barriers have been added to the fiber to solve this problem.

Silicon carbide (SiC) fiber

Boron fiber can react with metal matrices. This problem led to the development of an alternative fiber reinforcement—silicon carbide (SiC) fibers. Silicon carbide fiber has a tensile strength of 573,000 psi (3.95 GPa), which is higher than that of boron fiber. Its modulus of elasticity is similar to boron fiber at 58,000,000 psi (400 GPa). Metal matrix composites reinforced with silicon carbide can be easily produced. This is because silicon carbide fibers readily bond to the metal matrix and resist degradation during high-temperature processing. As a result, aluminum composites can be formed using simple high-temperature processes, such as investment casting (lost-wax casting) and low-pressure hot molding. When used in titanium composites, silicon carbide fibers can withstand extended exposure at diffusion-bonding temperatures without fiber degradation.

Carbon fiber

Carbon fiber is basically a long, thin filament of carbon with a diameter between 160 μinch and 430 μinch (4 μm and 11 μm). Nearly all commercial carbon fibers are manufactured by thermal charring, or "carbonizing," of an organic fiber. This is followed by a heat treatment to increase carbon content and achieve optimum physical and chemical properties. The final heat treatment temperature is between 1830°F and 3630°F (1000°C and 2000°C) for high-strength and ultrahigh-strength carbon fibers. A temperature between 3630°F and 5430°F (2000°C and 3000°C) is used for carbon fibers with high and ultrahigh modulus of elasticity. The term *graphite fiber* refers to carbon fibers that have been treated in the upper heat-treating temperature range. The term *carbon fiber* refers to carbon fibers treated at the lower heat-treating temperature range. Figure 15-2 shows typical properties for different types of carbon fibers.

Figure 15-2.
This table shows various properties of several carbon fiber types.

PROPERTIES OF CARBON FIBERS				
Fiber Type	Density, g/cm^3	Modulus of Elasticity	Tensile Strength	Electrical Resistivity, ohmAm
High-strength (PAN)	1.7–1.8	33–36 10^6 psi (228–248 GPa)	410–580 ksi (2.8–4.0 GPa)	12–30
Ultrahigh-strength (PAN)	1.7–1.8	38–42 10^6 psi (262–290 GPa)	590–830 ksi (4.1–5.7 GPa)	14–20
High-modulus (PAN, mesophase pitch)	1.8–2.0	50–80 10^6 psi (345–552 GPa)	250–500 ksi (1.7–3.4 GPa)	5–10
Ultrahigh-modulus (mesophase pitch)	2.0–2.2	90–130 10^6 psi (620–897 GPa)	300–360 ksi (2.1–2.5 GPa)	1–4
Low-modulus (rayon, pitch)	1.3–1.7	6–9 10^6 psi (41–62 GPa)	85–145 ksi (0.6–1.0 GPa)	30–100

Aluminum oxide fiber

Composites reinforced with aluminum oxide fibers are ideal for applications requiring light weight at elevated temperatures. Under these conditions, composites with aluminum oxide fibers are superior to unreinforced metals in stiffness, strength, fatigue performance, and wear characteristics. Aluminum oxide fibers can be used in a number of metal matrices, especially aluminum and magnesium.

Aluminum oxide is chemically inert. This reduces fiber degradation during high temperature fabrication and under working conditions. Aluminum oxide fibers also have good mechanical properties. Some types of fibers approach a tensile strength of 58,000,000 psi (400 GPa). Because their strength and stiffness remain high at elevated temperatures, the composites creep less at high temperature and have significantly better fatigue resistance than unreinforced metals. The hardness of aluminum oxide accounts for the exceptionally good wear resistance of composites with aluminum oxide reinforcement. However, the main shortcoming of aluminum oxide is its relatively high density (3.2 g/cm^3 to 4.0 g/cm^3).

Tungsten fiber

Tungsten is a refractory metal. A refractory metal is one which has a high melting point. Tungsten fibers increases strength, stiffness, and high-temperature capability of composites. It also improves ductility and thermal/electrical conductivity. Applications of tungsten-reinforced metal matrix composites are particularly appropriate for turbine blades, pressure vessels, flywheels, and simple loaded beams.

Tungsten fiber is also used in fiber reinforced superalloys. *Fiber reinforced superalloy (FRS)* refers to a class of engineering materials in which an oxidation-resistant alloy matrix is reinforced with strong, stiff, creep-resistant fibers. These composites are used for high-temperature applications.

DISCONTINUOUS FIBERS

Discontinuous fibers vary in length depending on the fiber manufacturing process. In general, short fibers are easily incorporated into a metal matrix.

Carbon fiber

Discontinuous carbon fibers consist of short filaments ranging from 0.008" (0.2 mm) to 1" (25 mm) in length. Uniform-length fibers are obtained by chopping continuous carbon yarns. Inexpensive, discontinuous fibers of inconsistent length are sometimes obtained by cutting or grinding random-length fibers. Chopped, high-modulus P55 and P100 type carbon fibers are the primary discontinuous reinforcements used. They are used with both aluminum and magnesium alloy matrices. These metal matrix composites are low density and have low thermal expansion.

Silicon carbide whisker

Whiskers are noncircular single crystals with a high aspect ratio. In other words, average length of a whisker is considerably greater than the average diameter. Most whiskers have aspect ratios ranging from about 50 to 150. Single-crystal whiskers usually have much greater tensile strengths than other types of discontinuous reinforcements.

Silicon carbide (SiC) whiskers have diameters ranging from 120 µinch to 390 µinch (3 µm to 10 µm) and lengths as great as 4" (100 mm). SiC whiskers produced by pyrolysis of rice hulls have an average diameter of 24 µinch (0.6 µm) and lengths ranging from 400 to 3000 µinch (10.2 to 76.2 µm). The tensile strength can approach 1,015,000 psi (7,000 MPa). Commercial quantities of rice-hull-derived SiC whiskers are routinely manufactured in the US for use in aluminum and magnesium matrix composites.

Refractory metal filaments

Refractory metal filaments are made from tungsten (W), niobium (Nb), or chromium (Cr). They are often incorporated into metal matrix by deformation processes. For example, copper and niobium have very limited solid solubility at room temperature. After melting and casting, a separate phase of niobium is formed in the copper matrix as it solidifies. Then, the cast ingot is subjected to extensive deformation, such as drawing and rolling. During the deformation, the niobium is broken into small filaments in the copper matrix. This is what results in a deformation-processed composite. Since the niobium does not dissolve into the copper matrix, the composite maintains high electrical and thermal conductivity while achieving high strength.

PARTICULATES

Particulates are used in metal matrix composites mainly to improve the strength of composites. The most common strengthening particulates are silicon carbide (SiC) and aluminum oxide (Al_2O_3). They are used in aluminum and copper matrix composites. For instance, a family of aluminum-silicon alloy matrix composites are commercially available. These composites contain up to 20% silicon carbide or aluminum oxides. The silicon-carbide-reinforced composites are used for gravity-casting and die-casting. The aluminum-oxide-reinforced composites are used for wrought materials. Aluminum oxide particulates have also been used to strengthen copper alloys without significant decrease in electrical and thermal conductivity.

Particulate-reinforced composites are manufactured economically in large batches by mixing the particles into the molten aluminum alloy. The molten mixture is then cast into ingots, extrusion billets, or rolling blooms. Products can be formed using conventional aluminum fabrication methods, such as casting, extrusion, rolling, and forging.

Another major purpose of particulates used in metal matrix composites is to serve as a solid lubricant. A typical application is graphite particulates used in an aluminum alloy matrix. The graphite works as a lubricant for components that may not be fully lubricated under certain conditions. This type of metal matrix composite is promising for applications such as bearings for aerospace and automobiles. These bearings may not receive full lubrication or may not be lubricated at all under service conditions.

MANUFACTURING

Metal matrix composites are often manufactured by traditional metal or alloy fabrication processes, such as casting, powder metallurgical processes, and deformation processes. Primarily, manufacturing methods of metal matrix composites can also be divided into three categories:

- ♦ casting and infiltration in which melting of the metal matrix is involved.
- ♦ powder metallurgical processes in which the matrix and reinforcement are not melted but bonded at high temperature.
- ♦ deformation processes in which the composite is deformed to form the desired product shapes and at the same time control the morphology and distribution of the reinforcement.

Complex shapes with selective composite reinforcement can be fabricated by innovative methods such as superplastic forming, diffusion bonding, or hot isostatic pressing. *Superplastic forming,* or diffusion bonding, uses materials that can be elongated or deformed almost indefinitely under suitable conditions without local fracturing. *Hot isostatic pressing (HIP)* forms the part under pressure from all directions and at high temperature in a pressure vessel.

CASTING AND INFILTRATION PROCESSES

Various techniques have been developed which involve at least partial melting of the matrix metal. This generally promotes good contact between the matrix and reinforcement and, therefore, a stronger bond. However, it can also lead to the formation of a brittle layer between the matrix and reinforcement that weakens the composite. These high-temperature interactions are sensitive to a number of factors, such as liquid/ceramic type, contact time, and pressure. There are three common types of casting or infiltration processes. These are slurry casting, squeeze casting, and infiltration.

Slurry casting

Slurry casting is a process in which liquid metal is simply stir-mixed with solid ceramic particles and the slurry solidifies on cooling. The slurry can be continuously agitated while ceramic particles are progressively added. The ceramic can be added to the matrix using conventional processing equipment. It can be carried out on a continuous or semi-continuous basis. This type of processing is now in commercial use for aluminum matrix composites reinforced by SiC particles.

Slurry mixture can be cast into molds by either conventional casting, die casting, or other pressure casting processes. Slurry casting is the simplest and most economical method of metal matrix composite manufacturing. However, there are some technical problems to be solved such as forming difficulties, inconsistencies in the microstructure, and interfacial chemical reactions.

Squeeze casting

The term *squeeze casting* refers to a process in which pressure is applied to a matrix that is solidifying. Usually, a single, hydraulically activated ram is used to apply the pressure. This technique generally results in a final product that is not very porous. This is due to efficient liquid feeding and compensation of shrinkage during solidification. A fine microstructure is also generated because of the rapid cooling induced by good thermal contact with a massive metal mold. These two features usually increase the strength of matrix alloys.

Unlike conventional pressure die casting, in squeeze casting the ram continues to move during solidification. This deforms the product's internal structure, compensating for the solidification contraction. In addition, the ram movement is usually slower and the applied pressure often is greater than that typical of die casting.

Squeeze casting can be applied to stir-mixed materials or powder mixtures. Aluminum matrix composites with continuous tungsten fibers have been produced through squeeze casting. The final products are not porous.

Infiltration

Liquid metal *infiltration* is a process of immersing fibers in a molten-metal bath to produce a metal matrix composite. The technique has been successfully used to fabricate copper-alloy matrix composites reinforced by continuous tungsten fibers.

In order to promote better infiltration of liquid metal into the fibers, high pressure is normally applied. This technique led to the development of *squeeze infiltration* in which liquid metal is injected into a preform. Figure 15-3 illustrates an apparatus used for squeeze infiltration. Squeeze infiltration has been particularly popular for very fine fibers, such as SiC whiskers. These fibers are difficult to blend with metallic powders and are impractical to stir-mix as well.

Another infiltration technique is *chemical vapor deposition (CVD) infiltration.* This technique is used to make fibers for producing graphite-aluminum and graphite-magnesium composites. Figure 15-4 shows an eight-line CVD infiltration unit. In CVD infiltration, graphite fibers are passed through a titanium boride vapor at a low temperature, often 1290°F (700°C). This results in an extremely thin layer of titanium boride being deposited on the fiber. The fiber is then wetted with molten aluminum or magnesium alloys to produce a graphite-reinforced composite wire. The composite wire is laid up into sheets, tubing, rods, and a variety of other shapes. CVD infiltration has also been modified and used to make woven graphite cloth, resulting in metal matrix composite sheets 12″ (300 mm) wide and up to 50′ (15 m) long. Other applications include surface metallization of monofilament or multifilament fibers.

Figure 15-3.
This is a schematic of a typical squeeze infiltration apparatus. Liquid metal is allowed to fall into the die cavity by withdrawing the sliding base of the crucible. The ram is then brought down to press the melt into the preform.

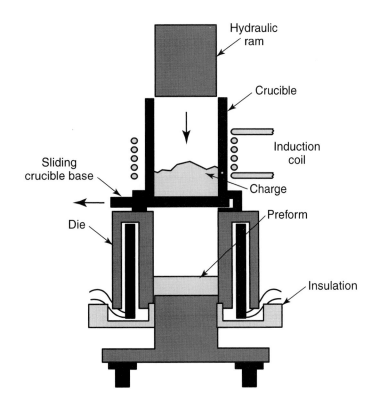

Figure 15-4.
A typical chemical vapor deposition (CVD) infiltration unit. The graphite filament, activated by the deposition of titanium boride film, is wetted by the aluminum and magnesium alloy to produce a graphite-reinforced composite wire. (ASM International)

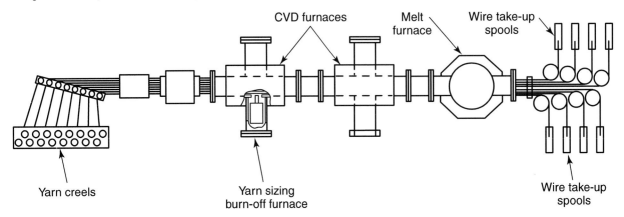

POWDER METALLURGICAL PROCESS

A metal matrix composite can be produced without melting the matrix. This is usually done with a powder metallurgical process. A *powder metallurgical process* involves compacting metal powders and sintering the compacted parts. This process is convenient and versatile for making metal matrix composites. It offers excellent control over the reinforcement content across a wide range. However, handling metallic powders requires careful following of safety procedures to prevent fire or explosion. Powdered aluminum and magnesium are reactive and flammable.

The process begins with mixing and blending alloyed metallic powder and ceramic whiskers, such as SiC whiskers. This is followed by heating and removing gas. Finally, the mixture is formed into the final part or stock material. Figure 15-5 shows a flowchart of a typical powder metallurgical process producing SiC-whisker-reinforced metal matrix composites.

High temperature is required to produce metal matrix composites from powders. The general sequence is:
1. Loading powder blends into a die assembly.
2. Placing the loaded die into the press and evacuating, heating, and removing gas to eliminate volatile contaminants.
3. Heating and pressing to consolidate the blend into a dense form.
4. Cooling the final part and removing it from the die.

Modification of powder metallurgical process has led to another technique called *diffusion bonding.* The process consists of hot-pressing an array of fibers between metal foils. Under elevated pressure and temperature, foils are deformed around the fibers. The foils are bonded to the fibers and to other foils. This process produces preforms that can then be laid up to form desired structures. Figure 15-6 shows a boron-aluminum composite fabricated using diffusion bonding. In addition to boron-aluminum composites, diffusion bonding is used for SiC-titanium and graphite-aluminum composites.

Figure 15-5.
This schematic shows the powder metallurgy process to produce SiC whisker-reinforced metal matrix composites.
(ASM International)

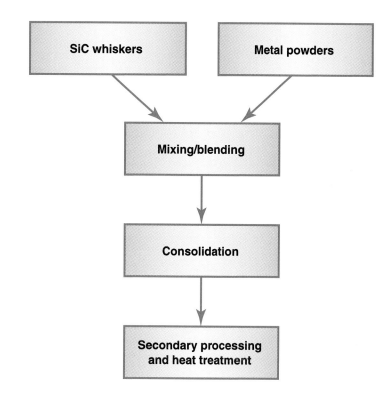

Figure 15-6.
This schematic shows a typical diffusion bonding process used to form boron-aluminum composites.
(ASM International)

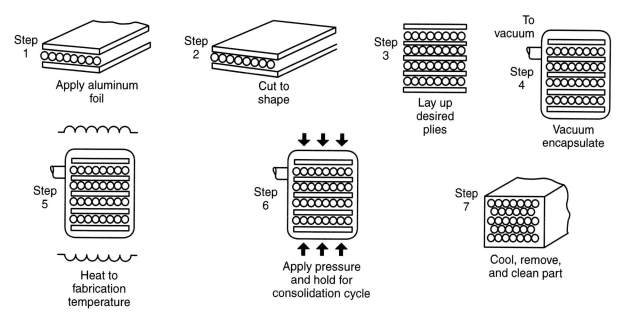

DEFORMATION PROCESS

One advantage of metal matrix composites is most metals are deformable. Therefore, a deformation process can be used to enhance composite properties and to achieve the designed shape. Commonly used deformation processes include extrusion, drawing, rolling, forging, and hot isostatic pressing. The main objectives of using deformation processes for metal matrix composites are to create a material that is not very porous, properly align fibers, and form the desired shape.

The deformation process has been used to produce copper-refractory-metal matrix composites. The refractory metals used include niobium or chromium. For example, copper and niobium are melted and then cast into an ingot. Niobium dendrites, or "fibers," form in the copper matrix after the mixture solidifies. The ingot is then subjected to extensive deformation by rolling or extrusion. This results in small fragments of niobium aligned with the deformation axis, as shown in Figure 15-7. Consequently, a deformation-processed, copper-niobium composite is produced. Since the copper matrix is not alloyed with niobium, electrical/thermal conductivity is not adversely affected. Moreover, niobium filaments increase the composite strength. As a result, the composite has a superior combination of mechanical strength and electrical/thermal conductivity.

PROPERTIES

Like polymer matrix composites, the properties of a metal matrix composite depends on the matrix, reinforcement, reinforcement distribution, and reinforcement orientation. This section presents some examples to illustrate typical mechanical, physical, and chemical properties of metal matrix composites.

MECHANICAL BEHAVIOR

Metal matrix composites generally offer the following advantages.
♦   Increased operation temperature.
♦   Strength and stiffness enhancement.
♦   Wear resistance enhancement.

Figure 15-7.
Extensive cold deformation of a Cu-Nb two phase solid can yield filaments of niobium aligned along the deformation direction.

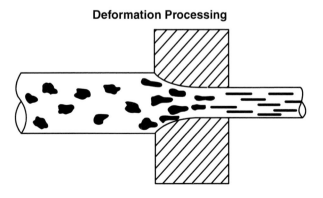

Deformation Processing

Higher operating temperature is the main advantage of metal matrix composites over polymer matrix composites and unreinforced alloys. For example, boron-aluminum composite retains good mechanical properties up to 950°F (510°C). However, an equivalent boron-epoxy composite is limited to only about 375°F (190°C). The strength and stiffness of metal matrix composites remain high at elevated temperatures. This results in less creep and significantly better fatigue resistance than unreinforced metals.

Metal matrix composites have higher strength and stiffness compared to unreinforced alloy metals and polymer matrix composites. For example, high-performance continuous SiC fiber in 6061 aluminum alloy yields a composite with a very high tensile strength of 200,000 psi (1380 MPa) and high modulus of elasticity of 30×10^6 psi (207 GPa). As a comparison, the base aluminum 6061 alloy has a tensile strength of 45,000 psi (310 MPa) and a modulus of elasticity of 8×10^6 psi (55 GPa).

The wear resistance of metal matrix composites is better than polymer matrix composites and unreinforced alloys. Saffil, a type of aluminum oxide fiber, has been used in selectively reinforced diesel engine pistons. The composite piston had about twice the wear and seizure resistance of the conventional iron piston.

Mechanical properties of fiber-reinforced metal matrix composites depend directly on the fiber orientation. Properties parallel to the fiber direction (longitudinal) are determined by the fiber. On the other hand, the properties perpendicular to the fiber direction (transverse) are determined by the matrix. Therefore, mechanical properties are very anisotropic. Because fibers dominate in the longitudinal direction, mechanical strength is generally quite high, whereas that in the transverse direction is much lower. As expected, the matrix alloy and heat treatment have little effect on the longitudinal properties but can cause a great deal of variation in the transverse direction.

PHYSICAL PROPERTIES

Density and electrical conductivity are important physical properties of metal matrix composites. These properties are important for light structural materials and electrical/thermal conductors.

The density of metal matrix composites follows the rule of mixture. That is, the density of any composite is proportional to the fraction and density of reinforcement and matrix. Figure 15-8 lists densities of selected reinforcements for metal matrix composites. Carbon and boron fibers have the advantage of being low in density, whereas silicon carbide has a little higher density. Refractory metals have higher density than most ceramic reinforcements.

Electrical conductivity is directly proportional to thermal conductivity. Both of these properties are significantly affected by alloying with standard metals. Since the matrix in a metal matrix composite does not alloy with the reinforcement, the conductivity of the matrix is not affected by the reinforcement. Therefore, metal matrix composites retain relatively high electrical/thermal conductivity while the reinforcement strengthens the composite. As a result, most metal matrix composites have a combination of high mechanical strength and high conductivity. For example, wires of Cu-18.2 vol.%Nb with a deformation strain of 11 reached a tensile strength 290,000 psi (2,000 MPa) whereas the conductivity still remains high at 40% IACS.

Figure 15-8.
This table shows the density of reinforcement in several metal matrix composites.

DENSITY OF REINFORCEMENTS FOR METAL MATRIX COMPOSITES	
Reinforcement	Density, g/cm³
Boron (B)	2.5
Silicon carbide (SiC)	3.045
Carbon (C)	1.7–2.2
Aluminum oxide (Al_2O_3)	3.2–4.0
Mullite	3.2
Aluminosilicate	2.7
Zirconia (ZrO_2)	5.7
Tungsten (W)	19.3
Niobium (Nb)	8.55
Chromium (Cr)	7.19

Figure 15-8.
This table shows the density of reinforcement in several metal matrix composites.

CHEMICAL PROPERTIES

The corrosion resistance of metal matrix composites is not affected significantly by the reinforcement because most reinforcements are chemically inert. Corrosion testing has been performed on aluminum 6061 alloy matrix composite reinforced with silicon carbide. The testing was conducted at the David W. Taylor Naval Ship Research and Development Center under various conditions, such as marine atmosphere, ocean splash/spray, alternate tidal immersion, and filtered seawater immersion. This composite performed well in all tests, exhibiting no more pitting damage than unreinforced aluminum 6061 alloy. In addition, a recent review on the corrosion behavior of aluminum-oxide-reinforced metal matrix composites concludes these composites do not have serious corrosion problems.

APPLICATIONS

Modern industrial machinery and equipment require materials of high strength, high stiffness, high creep resistance at elevated temperatures, high wear resistance, low density, high conductivity, and thermal stability. Metal matrix composites are developed to meet those demanding requirements. This section discusses several "real-life" examples to illustrate current and potential applications of metal matrix composites.

DIESEL ENGINE PISTON

Due to operating conditions, any diesel engine piston must retain strength at high temperature. The piston should also have good thermal conductivity, good thermal stability, and high wear resistance. Short aluminum oxide fibers reinforcing an aluminum matrix provides the desired properties for the piston. In addition, the piston is much lighter than conventional steel pistons. Figure 15-9 shows an example of a diesel engine piston made of aluminum matrix composite. The darker area shown is selectively reinforced by short aluminum oxide (Al_2O_3) fiber. The support region is unreinforced aluminum alloy.

This application represents one of the major successes in industrial use of metal matrix composites. Production of these pistons originated in Japan has been increasing steadily. Now, millions of units are produced each year. Originally, a nickel-cast-iron (Ni-resist7) insert was used in the piston ring area to prevent seizure of the piston ring with the top ring groove and bore. Unfortunately, this negatively affected heat flow and increased both weight and wear rate. Toyota Motor Company and Art Metal Manufacturing Company worked together to replace the cast iron insert with an aluminum composite reinforced with 5% aluminum oxide short fibers. The selective reinforcement was made possible by squeeze casting aluminum liquid into an alumina preform. A weight reduction of 5% to 10% was achieved. Test results showed that wear was reduced to less than 25% and seizure stress to 50% of the unreinforced aluminum alloy. The thermal conductivity of the insert is 400% better than the nickel-cast-iron insert.

Figure 15-9.
This schematic shows a typical application of an aluminum matrix composite reinforced with short aluminum oxide fibers. This application is a diesel engine piston.

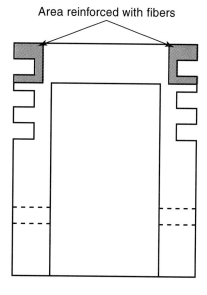

Area reinforced with fibers

BRAKE DRUM

Figure 15-10 depicts an automotive brake drum. The drum is high-pressure cast in DURALCAN® F3N.20S AL MMC. This composite contains virtually no Cu or Ni and is designed for corrosion-sensitive applications. The matrix is similar to 360 alloy and has high strength, high wear resistance, and good thermal properties.

Figure 15-10.
A—This automotive brake drum is high-pressure die-cast in DURALCAN® F3N.20S. This is a metal matrix composite containing virtually no Cu or Ni and is designed for use in corrosion-sensitive applications.
B—A microphotograph of the structure of DURALCAN®. (Duralcan USA)

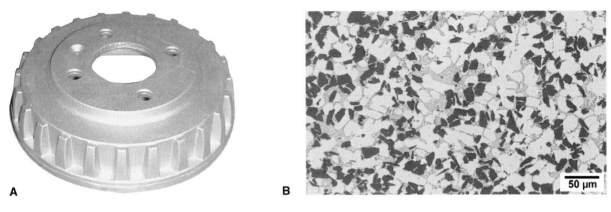

A   B

Figure 15-11.
This schematic shows the location of boron-aluminum struts and stabilizers used in the space shuttle. (ASM International)

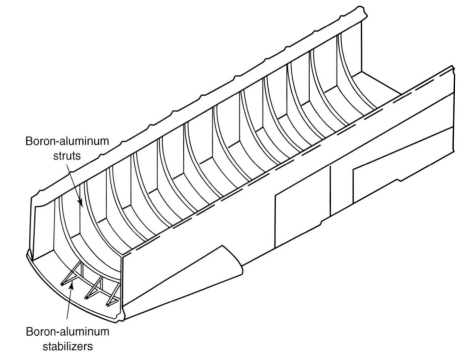

Boron-aluminum struts

Boron-aluminum stabilizers

SPACE SHUTTLE

Boron-aluminum composite is used for structural tubular struts, as frame and rib truss members in the mid-fuselage section, and as the landing gear drag link of the space shuttle orbiter. Figure 15-11 is a schematic of the mid-fuselage structure of the space shuttle showing boron-aluminum struts and stabilizers. The boron-aluminum composite application resulted in a 44% weight saving over the original aluminum design. This is because it is stronger and required less material to meet the strength criteria.

MICROELECTRONICS HEAT DISSIPATER

Boron-aluminum is currently available as a heat dissipater/cold plate material for multilayer board microchip carriers. This application makes use of the high thermal conductivity and low thermal expansion properties of boron-aluminum composite. Heat generated by closely packed semiconductor chips is removed by the boron-aluminum composite. Because boron-aluminum has a coefficient of thermal expansion close to that of chips, thermal stress due to the difference in thermal expansion between the chip and the substrate is significantly reduced. This fact greatly reduces the stress and fatigue of joints that hold chips to the substrate material. Therefore, the service reliability of electronic devices is significantly improved.

SUMMARY

Metal matrix composites have been developed to increase the working temperature, strength, and stiffness. They also improve wear resistance over polymeric matrix composite and conventional alloys and reduce the weight of some structural components. Material properties can be controlled and tailored to fit specific applications, such as strength, thermal conductivity, and thermal expansion.

A light metal or alloy is most frequently used as the matrix, except for electrical or thermal applications. These applications use silver or copper as the matrix.

The types and availability of reinforcements is always expanding. Basic types of reinforcement include continuous fiber, discontinuous fibers, short fibers, and particulates.

Metal matrix composites can be manufactured through casting, infiltration, powder metallurgy, deformation process, and many other innovative processes. There are many applications, including automobile, aerospace, biomedical, microelectronics, and many others.

IMPORTANT TERMS

Carbon Fiber

Chemical Vapor Deposition (CVD) Infiltration

Diffusion Bonding

Fiber Reinforced Superalloy (FRS)

Graphite Fiber

Hot Isostatic Pressing (HIP)

Infiltration

Metal Matrix Composite (MMC)

Powder Metallurgical Process

Slurry Casting

Squeeze Casting

Squeeze Infiltration

Superplastic Forming

QUESTIONS FOR REVIEW AND DISCUSSION

1. What is a metal matrix composite?
2. Why are aluminum matrix composites more popular than any other metal matrix composite?
3. What are the advantages and disadvantages of using titanium as a matrix over aluminum?
4. Which metal has the highest electrical conductivity—aluminum, titanium, copper, or silver?
5. What is the advantage of using silver as matrix for electrical contact material compared with copper?
6. What is the significance of boron fiber?
7. Why is a surface coating or diffusion barrier necessary for boron fibers used to reinforce metal matrix composites?
8. What is the major advantage of silicon carbide fiber over boron fiber?
9. What is the difference between carbon fiber and graphite fiber?
10. What is hot isostatic pressing (HIP)?
11. What is fiber reinforced superalloy?
12. What is a whisker?
13. What is slurry casting?
14. What is squeeze casting?
15. What is squeeze infiltration?
16. What is CVD infiltration?
17. What is diffusion bonding?
18. What are some benefits of using metal matrix composites?
19. What rule does the density of a metal matrix composite follow?

FURTHER READINGS

1. Clyne, T. W. and Withers, P. J. An Introduction to Metal Matrix Composites. Cambridge: Cambridge University Press (1993).
2. Everett, R. K. and Arsenault, R. J. Metal Matrix Composites: Mechanisms and Properties. Boston: Academic Press (1991).
3. ASM International. Engineering Materials Handbook: Composites. Materials Park, OH (1987).
4. Strong, A. B. Fundamentals of Composite Manufacturing: Materials, Methods, and Applications. Dearborn, MI: SME (1989).
5. Duralcan USA. DURALCAN Composites for High-Pressure Die Castings. San Diego, CA (1992).

INTERNET RESOURCES

www.mmm.com/market/industrial/mmc
   3M Metal Matrix Composites

www.lanxide.com
   Advanced Materials Lanxide

www.mmccinc.com
   Metal Matrix Cast Composites, Inc.

www.ultradiamondtech.com/applicationsmetalmatrixcomposites.html
   Ultra Diamond Technology, Inc.

www.dlcomposites.com/index.htm
   Honeywell Advanced Composites Inc.

www.ornl.gov
   Oak Ridge National Laboratory

www.msel.nist.gov
   NIST Materials Engineering and Science Lab

www.almmc.com
   The Aluminum MMC Consortium

www.alyn.com/index2.html
   Alyn Corporation

16

Introduction to Electronic Materials

KEY CONCEPTS

Upon completion of this chapter, you should understand:

➤ Conductivity and resistivity.
➤ The physical nature of conductors, insulators, and semiconductors.
➤ Conductivity and resistivity of materials within a family at different conditions.
➤ Ohm's Law.

The transistor, perhaps more than any other invention, ushered in the era of modern electronic devices. Its introduction in 1947 by Bell Laboratories and subsequent advancements in semiconductor electronics have truly moved the world into the Information Age. Data and images move from point to point, even around the world, with tremendous speed and clarity. This chapter explores the various materials used in the electronics industry. The properties that make them the appropriate choices for simple conductors to complex semiconductors or superconductors are also examined.

INTRODUCTION

At the most basic level, materials are selected for electrical and electronic applications for one of three reasons—they conduct, insulate, or are semiconductors. In order to clearly understand the origin of specific properties of any material, it is important to understand their atomic structure. Material properties are a result of their basic atomic structure. Basic atomic structure is discussed in detail in Chapter 2. However, the brief review presented here will provide a foundation for exploring the electronic properties of materials.

Electrical conductors are made from metallic elements. Metals readily conduct electrical and thermal energy because of their unique atomic structures and atomic bonds. Atoms of metals are bonded to one another by what is referred to as the metallic bond. Most metallic elements have one, two, or three valence electrons. In an attempt to reach stable electron configuration, metals "give up" their valence electrons to form what is referred to as a *sea of electrons* or *electron cloud,* Figure 16-1. This sea of electrons is shared by all atoms in the metallic mass. The movement of electrons freely through metals is the reason they readily conduct both thermal and electrical energy.

Insulators resist the flow of thermal and electrical energy. They are formed through ionic bonding or covalent bonding. Covalent bonds are extremely strong because they are formed by the sharing of electrons. Ionic bonds are also strong because they donate electrons, creating a stable electron configuration which resists energy flow.

Semiconductors are made of elements that border the metals and nonmetals on the periodic table of elements, Figure 16-2. Silicon (Si) and Germanium (Ge) are the most commonly used semiconducting materials. Their atomic bonds and properties are similar to those of ceramics. Only by adding impurities, called *dopants,* to these elements and compounds do semiconductors exhibit desired electrical properties.

ENERGY LEVELS AND ENERGY BANDS

To understand how materials conduct or resist the flow of electricity, it is important to understand the concept of energy levels and energy bands. Metals only have 1, 2, or 3 electrons in their valence shell; therefore, they give away or "donate" valence electrons. Metals with fewer valence electrons tend to conduct electricity better. Bonded metals have atoms closely packed in orbitals that overlap one another. This allows electrons to easily "jump" from the orbital of one atom to that of another. The free electrons, however, cannot all have the same energy. According to *Pauli's Exclusion,* each quantum state can only be occupied by two electrons having opposite spin numbers. This creates what is referred to as a *band of energy levels.*

Figure 16-1.
Metals tend to give up electrons to a "Sea of Electrons."

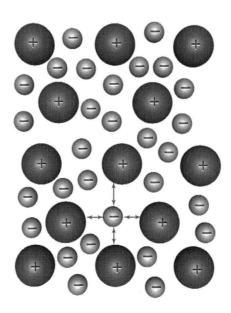

Figure 16-2.
Semiconductor elements are highlighted in this periodic table of elements.

Key to Chart:
- 26 — Atomic Number
- Fe — Element Symbol
- 55.847 — Atomic Weight

1 IA	2 IIA	3 IIIA	4 IVA	5 VA	6 VIA	7 VIIA	8 VIIIA	9 VIIIA	10	11 IB	12 IIB	13 IIIB	14 IVB	15 VB	16 VIB	17 VIIB	18 VIIIB
1 H 1.008																	2 He 4.003
3 Li 6.941	4 Be 9.0122											5 B 10.811	6 C 12.011	7 N 14.007	8 O 15.999	9 F 18.998	10 Ne 20.179
11 Na 22.989	12 Mg 24.305											13 Al 26.982	14 Si 28.086	15 P 30.974	16 S 32.066	17 Cl 35.45	18 Ar 39.948
19 K 39.098	20 Ca 40.078	21 Sc 44.956	22 Ti 47.88	23 V 50.942	24 Cr 51.996	25 Mn 54.938	26 Fe 55.847	27 Co 58.933	28 Ni 58.69	29 Cu 63.546	30 Zn 65.39	31 Ga 69.72	32 Ge 72.59	33 As 74.922	34 Se 78.96	35 Br 79.904	36 Kr 83.80
37 Rb 85.468	38 Sr 87.62	39 Y 88.906	40 Zr 91.224	41 Nb 92.906	42 Mo 95.94	43 Tc (98)	44 Ru 101.07	45 Rh 102.906	46 Pd 106.42	47 Ag 107.868	48 Cd 112.41	49 In 114.82	50 Sn 118.710	51 Sb 121.75	52 Te 127.60	53 I 126.905	54 Xe 131.29
55 Cs 132.905	56 Ba 137.33	57 La 138.906	72 Hf 178.49	73 Ta 180.948	74 W 183.85	75 Re 186.207	76 Os 190.2	77 Ir 192.22	78 Pt 195.08	79 Au 196.967	80 Hg 200.59	81 Tl 204.383	82 Pb 207.2	83 Bi 208.980	84 Po (209)	85 At (210)	86 Rn (222)
87 Fr (223)	88 Ra 226.025	89 Ac 227.028	104 Rf (261)	105 Db (262)	106 Sg (263)	107 Bh (262)	108 Hs (265)	109 Mt (266)	110 (269)	111 (272)	112 (277)						

*	58 Ce 140.12	59 Pr 140.908	60 Nd 144.24	61 Pm (145)	62 Sm 150.36	63 Eu 151.96	64 Gd 157.25	65 Tb 158.925	66 Dy 162.50	67 Ho 164.931	68 Er 167.26	69 Tm 168.934	70 Yb 173.04	71 Lu 174.967
†	90 Th 232.038	91 Pa 231.036	92 U 238.029	93 Np 237.048	94 Pu (244)	95 Am (243)	96 Cm (247)	97 Bk (247)	98 Cf (251)	99 Es (252)	100 Fm (257)	101 Md (258)	102 No (259)	103 Lr (260)

Each electron shell is capable of holding a specific number of electrons. Each shell, or orbital, has sublevels to accommodate the electrons. Sublevels are designated using the letters s, p, d, and f.

In covalently bonded elements (Group IV), the central atom shares a total of eight electrons and has the electron configuration s^2p^6. The s and p states are fully occupied and there are no empty states. Since there are no empty states, electrons cannot move when an electron field is applied to the materials. Therefore, materials with covalent bonds (Group IV) generally do not conduct electricity. They include ceramics and organic polymers.

Covalently bonded materials can, however, be made to conduct electric current by freeing electrons from the bond. This creates a vacant electron site called a *hole.* The atom also has a positive charge. These electrons can now conduct and are at a much higher energy state. The freed electrons are different from the electrons in the covalent bond. They are identified as being in a *conduction band,* rather than in a valence band.

OHM'S LAW

Most electrical properties are based on Ohm's Law. *Ohm's Law* states that voltage (V) is a product of current and resistance. Current is measured in amperes (I). Resistance (R) is measured in ohms (Ω). This is expressed in one of three ways:

$$V = IR$$

or

$$I = \frac{V}{R}$$

or

$$R = \frac{V}{I}$$

CONDUCTIVITY

Conductivity is a measure of a material's ability to conduct electrical current. It is often defined as the reciprocal of resistivity. Figure 16-3 shows the conductivity and resistivity of selected common materials. Conductivity is calculated:

$$\sigma = \frac{1}{\rho} = nq\mu$$

where σ is conductivity, ρ is resistivity, n is density of charge carriers (m^{-3}), q is the charge carried by each charge carrier (coulombs), and μ is the mobility of each carrier ($m^2/v(s)$).

Figure 16-3.
This table shows the conductivity and resistivity of various materials.

CONDUCTIVITY AND RESISTIVITY OF SELECTED MATERIALS			
Material	Electrical Conductivity $s(W-cm)^{-1}$	Thermal Conductivity (BTU/Sq.Ft./Hr/°F/In)	Electrical Resistivity Ohm/m
Copper	6.0×10^5	2500	1.7×10^{-8}
Carbon Steel (SAE 1020)	1.0×10^5	360	7×10^{-7}
Aluminum	3.8×10^5	1190	2.9×10^{-8}
Magnesium Alloy	2.2×10^5	522	4.45×10^{-8}
Titanium	—	108	42×10^{-8}
Nickel	1.5×10^5	306	6.84×10^{-8}
Gray Iron	1.4×10^4	310	9.71×10^{-8}
Silver	—	244	1.59×10^{-8}
Stainless Steel	1.4×10^4	113	16×10^{-9}
Polyethylene	$<10^{-16}$	.290	10^{15}
Nylon	10^{-12}	—	10^{13}
Polystyrene	$<10^{-16}$	.073	10^{16}
Silicon	4.3×10^{-4}	—	10^{13}
Al_2O_3	$<10^{-14}$	27	10^{12}
SiC	1.0×10^{-1}	109	.1

RESISTIVITY

Resistivity is a measure of how well a material resists current flow. Resistivity is a product of the material's microstructure and not of the size and length of the conductor. Do not confuse resistivity with resistance. Resistance is a specific measure based on physical dimensions. Resistivity is calculated:

$\rho = R(A/l)$

where ρ is resistivity, R is resistance, A is the cross sectional area of the conductor, and l is the length of the conductor.

CONDUCTORS

Conductors are materials with a magnitude of conductivity on the order of $10 \times 10^6 (\Omega \cdot cm)^{-1}$. Conductivity can be affected by environment and impurities. As temperature increases, electron mobility decreases. Also, even small amounts of impurities can dramatically reduce the conductivity of metals. For example, 0.2% phosphorous in copper can have a 20% loss of conductivity over pure copper.

The best conductors of electricity are the metals gold, silver, aluminum, copper, and their alloys. These metals make the best conductors of electricity because of their molecular structure. They all have face centered cubic (FCC) lattice structure and have only a few valence electrons.

INSULATORS

Insulators are nonconducting materials. In other words, they are materials with low conductivity, Figure 16-3. Structurally, these materials are ionically or covalently bonded with energy gaps much larger than conducting materials. The ionic and covalent bonds restrict electron and ion mobility. As a result, electrons do not readily move from one energy level to another, making them good insulators. No material is a perfect insulator.

Approximately 80% of industrial ceramics are insulators. Most ceramics are electrical insulators and are widely used for insulating electric power lines. Glass and porcelain are the most common line insulators, Figure 16-4. Not only do they resist the flow of electricity, but they are inexpensive. In addition to power line insulators, ceramics are also used in the electronics components industry. They are commonly used as substrates. A substrate is the structural, insulated base on which electrical components are placed.

Figure 16-4.
These electrical line insulators are made from glass.

At one point in history, wire insulation was made of natural rubber. Today, wire insulation is made of plastics and synthetic rubber. Requirements for high thermal conductivity in electrical components has led to the use of beryllia, high-purity alumina, and aluminum nitride. Materials that make excellent insulators are the pure metal oxides (Al_2O_3, MgO), silicate ceramics, and organic polymers (vinyl polymers, polyolefins).

SEMICONDUCTORS

Semiconductors are materials with conductivity somewhere between a conductor and an insulator. Relatively few elements are used in the manufacture of semiconductors, Figure 16-5. Silicon is perhaps the most common of these elements. With both positive and negative charge carriers present in moderate numbers, silicon demonstrates a moderate value of electrical conductivity between conductors and insulators. Silicon is also extremely abundant and inexpensive.

Semiconducting elements are covalently bonded solids that appear on the periodic table of elements on the border line that separates metals and nonmetals. They are generally neither good conductors nor good insulators. The most commonly used semiconductor elements and compounds are silicon (Si), germanium (Ge), and gallium arsenic (GaAs). Semiconductors are typically used in rectifiers to change alternating current to direct current. Semiconductors are also used in transistors to amplify current. Other applications include converting heat energy to electric energy, such as a solar cell.

Semiconductors made from a pure element are called *intrinsic semiconductors.* Silicon is an intrinsic semiconductor. The term intrinsic means the flow of electrons is a product of increasing the temperature, rather than adding impurities to the semiconducting material. Semiconductors with carefully controlled small amounts of impurities, or dopants, are called *extrinsic semiconductors.*

The number of electrons in the conduction band can be greatly increased by adding dopants from Group V of the periodic table of elements, Figure 16-5. These elements are referred to as *donor atoms.* Phosphorous and arsenic are commonly used as dopants in extrinsic semiconductors. Phosphorous, for example, has five valence electrons. This is one more than silicon. A small increase in thermal energy can dislodge the "donor" electron from phosphorous, giving silicon one additional electron and a negative charge. The conduction mechanism in semiconductors with donor atoms is dominated by negative charge carriers. These doped semiconductors are called *n-type* semiconductors.

Some dopants have one less electron in their valence shell than in the valence shell of the semiconducting material. These dopants come from Group III in the periodic table of elements, Figure 16-5. They include aluminum (Al), Indium(In), Gallium (Ga), and Boron (B). These elements, referred to as *acceptor impurities,* tend to attract electrons from the silicon atom, thus creating a positively charged silicon ion. The conduction mechanism in these semiconductors is dominated by the positive charge carriers and are known as *p-type* semiconductors.

Figure 16-5.
Group III and V are shown highlighted in this periodic table of elements.

Key to Chart

- 26 — Atomic Number
- Fe — Element Symbol
- 55.847 — Atomic Weight

1 IA	2 IIA	3 IIIA	4 IVA	5 VA	6 VIA	7 VIIA	8 VIIIA	9 VIIIA	10 VIIIA	11 IB	12 IIB	13 IIIB	14 IVB	15 VB	16 VIB	17 VIIB	18 VIIIB
1 H 1.008																	2 He 4.003
3 Li 6.941	4 Be 9.0122											5 B 10.811	6 C 12.011	7 N 14.007	8 O 15.999	9 F 18.998	10 Ne 20.179
11 Na 22.989	12 Mg 24.305											13 Al 26.982	14 Si 28.086	15 P 30.974	16 S 32.066	17 Cl 35.45	18 Ar 39.948
19 K 39.098	20 Ca 40.078	21 Sc 44.956	22 Ti 47.88	23 V 50.942	24 Cr 51.996	25 Mn 54.938	26 Fe 55.847	27 Co 58.933	28 Ni 58.69	29 Cu 63.546	30 Zn 65.39	31 Ga 69.72	32 Ge 72.59	33 As 74.922	34 Se 78.96	35 Br 79.904	36 Kr 83.80
37 Rb 85.468	38 Sr 87.62	39 Y 88.906	40 Zr 91.224	41 Nb 92.906	42 Mo 95.94	43 Tc (98)	44 Ru 101.07	45 Rh 102.906	46 Pd 106.42	47 Ag 107.868	48 Cd 112.41	49 In 114.82	50 Sn 118.710	51 Sb 121.75	52 Te 127.60	53 I 126.905	54 Xe 131.29
55 Cs 132.905	56 Ba 137.33	57 La 138.906 *	72 Hf 178.49	73 Ta 180.948	74 W 183.85	75 Re 186.207	76 Os 190.2	77 Ir 192.22	78 Pt 195.08	79 Au 196.967	80 Hg 200.59	81 Tl 204.383	82 Pb 207.2	83 Bi 208.980	84 Po (209)	85 At (210)	86 Rn (222)
87 Fr (223)	88 Ra 226.025	89 Ac 227.028 †	104 Rf (261)	105 Db (262)	106 Sg (263)	107 Bh (262)	108 Hs (265)	109 Mt (266)	110 (269)	111 (272)	112 (277)						

*	58 Ce 140.12	59 Pr 140.908	60 Nd 144.24	61 Pm (145)	62 Sm 150.36	63 Eu 151.96	64 Gd 157.25	65 Tb 158.925	66 Dy 162.50	67 Ho 164.931	68 Er 167.26	69 Tm 168.934	70 Yb 173.04	71 Lu 174.967
†	90 Th 232.038	91 Pa 231.036	92 U 238.029	93 Np 237.048	94 Pu (244)	95 Am (243)	96 Cm (247)	97 Bk (247)	98 Cf (251)	99 Es (252)	100 Fm (257)	101 Md (258)	102 No (259)	103 Lr (260)

In semiconductor materials, the valence electrons are normally stable. However, when a dopant is added to the semiconductor, the energy in the material is raised or lowered. This causes the displacement of valence electrons, Figure 16-6. As the electrons jump energy levels, holes are left. This creates an imbalance. For example, atoms of silicon have four valence electrons in the outer shell. They bond to each other covalently by sharing electrons. When the dopant boron is added, with three electrons in its valence shell, it tends to attract an electron from one of the silicon atoms. This causes a hole and creates an electron imbalance. A p-type semiconductor is produced because there is one more proton than electrons.

When a doping element with five electrons in its outer shell is added to a silicon crystal, it tends to release electrons. This creates an n-type semiconductor. Bonding an n-type and a p-type together creates a *p-n junction*, Figure 16-7. A negative voltage applied to the p side causes the electrons of

Figure 16-6.
When a dopant is added to the semiconductor, the energy in the material is raised or lowered.

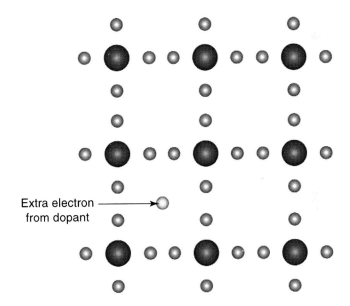

Extra electron from dopant

Figure 16-7.
This schematic shows a p-n junction in a circuit.

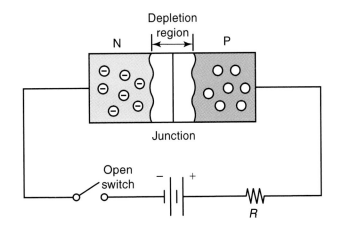

the three valence atoms to move away from the junction. The crystal then resists electrical flow. However, when voltage is applied to the n side, electrons are pushed across the junction. This causes electrons to flow. A p-n junction is used to create a diode or a rectifier. A rectifier is used to change alternating current (ac) into pulsating direct current (dc). When the semiconductors are assembled in three layers, such as a *p-n-p junction* or *n-p-n junction,* a weak current applied to the middle wafer increases the flow of electrons across the whole unit. This forms a transistor.

SUPERCONDUCTORS

As stated before, resistivity in materials is influenced by temperature. Resistivity increases as the temperature increases and decreases as the temperature decreases. In other words, a material is a better conductor at lower temperatures. Generally, however, there is a bottom temperature limit. Cooling the material below this temperature does not result in an increase in conductivity. There are, however, some exceptions.

At a critical temperature (T_c), the resistivity of mercury (Hg) drops to zero and it becomes a perfect conductor. A perfect conductor is called a *superconductor.* Approximately 25 elements, a number of compounds, and many alloys are known to exhibit superconductivity. Niobium, vanadium, lead, and their alloys are examples of metals that can become superconductors below critical temperature. The transition temperature to superconductivity varies by material. The critical temperature range for these materials is .01° Kelvin to 125° Kelvin.

The compound $YBa_2Cu_3O_7$ has been one of the most widely studied superconductors. Metals exhibiting superconductivity at near absolute zero include iridium, lead, mercury, columbium, tin, tantalum, vanadium, and many alloys. Commercially available alloys considered among the best superconductors are lead-molybdenum-sulfur, columbium-tin, and columbium-titanium.

The problem with current superconductors is that they function at extremely low temperatures. The ultimate goal of current research is to create a room-temperature superconducting material. Recently, alloys and compounds that exhibit superconductivity at temperatures well above absolute zero have been discovered. Among these are compounds of lanthanum, strontium, copper, and oxygen. These are superconductive at –400°F (–240°C). There is also a barium-yttrium-copper oxide which is superconductive at –296°F (–183°C).

Potential uses for superconductors include powerful and efficient electromagnets for levitated trains. These trains travel above the electromagnetic tracks, thus eliminating friction. Electromagnets are also used in magnetic resonance imaging devices (MRIs). These medical devices are used to diagnose tumors and physical problems with the brain, among other things.

SUMMARY

The invention of the transistor ushered in the era of modern electronics. This eventually lead to the Information Age. As a result of quick and continued advancements in electronic equipment, new and improved materials have been required. Conductors, insulators, and semiconductors all play an important role in any modern electronic device.

There is also a need for materials that conduct current with no loss of energy. Research into superconductors continues to be the focus of many worldwide efforts. Current superconductors operate at low temperatures. The search continues for a room-temperature superconducting material.

IMPORTANT TERMS

Acceptor Impurities

Band of Energy Levels

Conduction Band

Conductivity

Donor Atoms

Dopants

Electron Cloud

Extrinsic Semiconductors

Hole

Insulators

Intrinsic Semiconductors

N-P-N Junction

N-Type

Ohm's Law

Pauli's Exclusion

P-N Junction

P-N-P Junction

P-Type

Sea of Electrons

Superconductor

QUESTIONS FOR REVIEW AND DISCUSSION

1. Why are metals better conductors of electrical and thermal energy than other materials?
2. What is the definition of conductivity?
3. What is the difference between resistance and resistivity ?
4. How does temperature affect the electrical properties of materials?
5. Which materials are perfect insulators?
6. List some examples of materials used as insulators.
7. Why is silicon the most commonly used semiconductor material?
8. What is an intrinsic semiconductor?
9. What is an extrinsic semiconductor?
10. What functions do elements such as phosphorous and arsenic have in the creation of semiconductors?
11. What are dopants and what is their function?
12. What is a transistor?
13. What are superconductors?

FURTHER READINGS

1. Bauccio, Michael (ed). Engineering Materials Handbook. Materials Park, OH: ASM International (1994).
2. Fink, Donald G. and Christiansen, Donald (Eds). Electronics Engineer's Handbook 3rd ed. New York: McGraw-Hill Inc. Company, (1997).
3. Gibilisco, Stan (ed.). The Illustrated Dictionary of Electronics, 7th ed. New York: McGraw-Hill Inc. (1997).
4. Parker, Sybil P. (ed). Encyclopedia of Electronics and Computers, 2nd ed. New York: McGraw-Hill Book Inc. (1987).

INTERNET RESOURCES

www.ieee.org
Institute for Electrical and Electronic Engineers

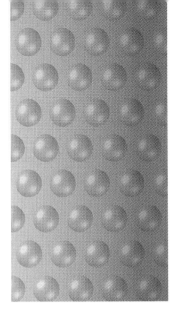

Chapter 17

Introduction to Standards and Nondestructive Testing of Industrial Materials

KEY CONCEPTS

Upon completion of this chapter, you should understand:
- ➤ The purpose and applications of industrial materials testing.
- ➤ The need for materials testing standards and standardizing agencies.
- ➤ Five common nondestructive tests.
- ➤ Principles and applications of common nondestructive tests.

The purpose of testing any material is to understand the properties of the material. All areas of industry, especially those dealing with structures and machines, are concerned with material properties. Materials must be tested for several reasons including safety, product performance, product quality, new product and process development, and scientific research.

Certain standards have been developed to aid in testing. In addition many testing methods exists. These include both destructive and nondestructive tests. Destructive testing is discussed in Chapter 18. This chapter introduces testing standards and standardizing agencies. Then, commonly used nondestructive testing methods are discussed.

INTRODUCTION TO STANDARDS

As stated above, there are several reasons that a material is tested. These reasons are listed and briefly explained below.

- ◆ **Safety and product performance.** All structures and machinery are designed to support specified loads and provide certain services. That translates into certain material property requirements, especially mechanical properties, to ensure the structure and machinery meet the performance and safety requirements.
- ◆ **Product quality.** The variation in material property and product performance must be known and kept to a minimum. This helps maintain the product quality level and the profitability of manufacturing process.

- ◆ **New product and process development.** In a competitive world, developing new products and processes is essential to maintain leadership in the market. Material properties must be tested to understand and develop new products or improve existing products.
- ◆ **Scientific research.** Modern industry has created the need for new and better materials. To develop new materials and improve existing materials by different processes, material properties must be tested and understood.

Standardized tests are followed for almost all industrial material testing applications. It is very important to follow standard tests because different test methods and procedures usually produce different results. In other words, you must make sure you are always comparing "apples to apples." Standards provide a common base of comparison for various test results. Standards also offer a common "language" for communication between different organizations in industry and commerce.

Industrial standards regarding material specifications and material testing are generated and maintained by various agencies. These include: standardizing organizations, such as the American Society for Testing and Materials (ASTM); professional or industrial associations, such as the Society of Automotive Engineers (SAE); and governmental agencies, such as the Department of Defense.

AMERICAN SOCIETY FOR TESTING AND MATERIALS (ASTM)

Founded in 1898, *American Society for Testing and Materials (ASTM)* is a scientific and technical organization. It was formed for "the development of standards on characteristics and performance of materials, products, systems and services and the promotion of related knowledge." ASTM is the world's largest source of voluntary standards. The society operates through 135 main technical committees and 2,080 subcommittees. These committees function in prescribed fields under regulations that ensure balanced representation among producers, users, and general interest participants. ASTM standards cover almost all industrial materials, including metals and alloys. ASTM standards also cover construction products, petroleum products, paints and coatings, textiles, plastics, rubber, electrical insulation and electronics, water technology, nuclear, solar and geothermal energy, and medical devices. Most material test standards are maintained by ASTM.

ASTM publishes a collection of its standards called the *Annual Book of ASTM Standards.* This contains periodically revised, formally approved ASTM standard classifications, guides, practices, specifications, test methods, terminology, and related material. ASTM produces six principal types of standards, as described below.

- ◆ **Standard test method.** A definitive procedure for identification, measurement, and evaluation of one or more qualities, characteristics, or properties of a material, product, system, or service that produces a test result.

- ◆ **Standard specification.** A precise statement of a set of requirements to be satisfied by a material, product, system, or service. A specification also recommends procedures for determining whether each requirement is satisfied.
- ◆ **Standard practice.** A definitive procedure for performing one or more specific operations or functions that do not produce a test result.
- ◆ **Standard terminology.** A document containing terms, definitions, descriptions of terms, explanations of symbols, abbreviations, or acronyms.
- ◆ **Standard guide.** A series of options or instructions that do not recommend a specific course of action.
- ◆ **Standard classification.** A systematic arrangement or division of materials, products, systems, or services based on similar characteristics. Similar characteristics may include material origin, composition, properties, or use.

AMERICAN NATIONAL STANDARDS INSTITUTE (ANSI)

When the American National Standards Institute (ANSI) was founded in 1918, standardization activities were just beginning in the United States. Many groups were developing standards. Sometimes the interests and activities of these groups overlapped and conflicted. The result was a waste of time and money, and considerable confusion. Five professional and technical societies and three government departments decided a single organization was needed to coordinate everything. As a result, ANSI was created to handle the task.

The American National Standard Institute coordinates standard activities on the national level. It is a federation of standards from commerce, industry, government, and professional, trade, consumer, and other organizations. The major goals and activities of ANSI include identifying needs for standards, protecting public interests, protecting consumer rights, promoting standard development and implementation, approving American national standards, and representing interests of the United States in nongovernmental international standards work.

PROFESSIONAL ASSOCIATIONS AND GOVERNMENTAL AGENCIES

Many professional and trade associations and some governmental agencies are actively involved in standards related to materials used for industry. Examples include the American Iron and Steel Institute, Society of Automotive Engineers, American Concrete Institute, Society of Plastics Industry, National Institute of Standards and Technology, Department of Defense, and the Food and Drug Administration.

American Iron and Steel Institute (AISI)

The American Iron and Steel Institute (AISI) is a trade association that governs materials specification standards on iron and steel. Its goals are: to provide high-quality, value-added products to a wide array of customers; to lead the world in innovation and technology in the production of steel; to produce steel in a safe and environmentally friendly manner; and to increase the market for North American steel in both traditional and innovative applications.

Society of Automotive Engineers (SAE)

The Society of Automotive Engineers (SAE) promotes structural material applications in automotive and aerospace industries. Many standards have been established related to structural steels and steel alloys.

American Concrete Institute (ACI)

The American Concrete Institute (ACI) recommends standards regarding cement, materials made from cement, and applications of cement. It is dedicated to improving the design, construction, manufacturing, and maintenance of concrete structures and facilities. ACI develops and disseminates information and standards on concrete uses.

Society of Plastics Industry (SPI)

The Society of Plastics Industry (SPI) is an important trade association dealing with polymeric materials. SPI promotes proper uses of plastics by establishing standards, test procedures, and specifications for plastic pipe, bottle, furniture, packaging, and various resins.

National Institute of Standards and Technology (NIST)

Formerly known as National Bureau of Standards (NBS), the National Institute of Standards and Technology (NIST) has an overall goal of strengthening and advancing the nation's science and technology and facilitating their effective application for public benefit. NIST conducts research to provide a basis for the nation's physical measurement system, and offers scientific and technological services for industry and government. NIST promotes competitiveness for industry, maintains equity in trade and technical services, and promotes public safety.

Department of Defense (DOD)

The Department of Defense (DOD) is a large governmental agency dealing with military applications of materials. It is made up of various agencies with widely varied functions. However, some of these agencies are charged with maintaining various military standards on materials, processes, and products to ensure the quality and performance of military equipment and supply.

Food and Drug Administration (FDA)

The Food and Drug Administration (FDA) makes policies aimed at protecting the health of the nation against impure and unsafe foods, drugs, cosmetics, and other potential hazards. Implications for material scientists and engineers are related to processing equipment, containers, medical devices, and implants.

NONDESTRUCTIVE TESTING

Many material tests result in damage or failure of the test piece. This type of test is called destructive testing and is discussed in Chapter 18. However, it is often desirable to know properties of a material or product without subjecting it to destructive testing. A material test that does not result in damage or failure of the test piece is called a ***nondestructive test.***

Advantages of using nondestructive tests are very straight forward since the part is not damaged or destroyed. Usually, nondestructive testing is faster than destructive testing because it does not need extensive specimen preparation. Therefore, it can be used for production online control or construction site inspection. A nondestructive test is also used to detect any flaws existing in raw materials or semifinished products before any further machining or fabrication is performed. Therefore, nondestructive tests can be used for quality control and production cost saving.

There are many methods of nondestructive testing and many commercial instruments available. The next section briefly describes five commonly used nondestructive material tests.

PENETRANT TEST

Penetrant tests are a simple method to detect defects with surface openings. Figure 17-1 illustrates the principal of penetrant method. In this method, the surface is first cleaned and dried. A penetrating dye fluid is then sprayed or swabbed onto the surface. The part is allowed to stand for

Figure 17-1.
A—The penetrant test starts by coating the cleaned surface with penetrant, which seeps into any cracks. B—Excess penetrant is removed from the surface before the developer is applied. C—Surface cracks are indicated by a small amount of penetrant drawn out by the developer.

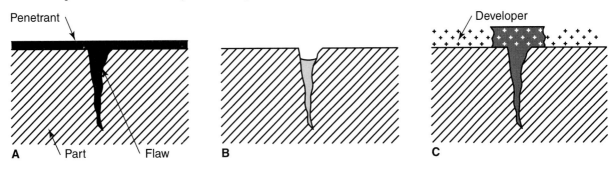

sufficient time for the penetrant to fill in any surface cracks. The excess penetrant is flushed from the surface with warm water or cleaner. After that, the surface is carefully dried and coated with developer. The dye left in surface or subsurface cracks is drawn to the developer, which shows the crack on the surface. After suitable developing time, the parts can be inspected for defects exposed on the surface.

When visible dyes, such as red dyes, are used, the parts are inspected under normal white light. The operator looks for red dye contrasting against a white developer background. When fluorescent dyes are used, parts are inspected under a "black light." The operator looks for a bright yellow-green color against a deep blue-violet background. In both cases, the operators are looking for small amounts of penetrating dye that show the actual flaws.

RADIOGRAPHIC TEST

Radiographic tests are similar to dental X ray inspection. It uses short wavelengths, such as X rays, beta rays, and gamma rays. Figure 17-2 shows the principal of a radiographic test using X rays. X rays can penetrate a material to certain level, according to its internal structure of the material. A film is placed behind the part and exposed to the radiation source penetrating the material. The film is developed and shows an image of the internal structure of the material. This test method is useful for detecting flaws in welded joints, especially for critical structures.

In operation of any radiographic instrument, safety precautions must be taken to avoid accidents. All radiographers and radiographic assistants must wear safety shielding. They must also carry an individual direct reading pocket dosimeter and either a film badge or a thermoluminescent dosimeter. Overexposure to X rays, beta rays, and gamma rays can lead to injuries or health problems.

Figure 17-2.
In the radiographic method, a film is placed behind the parts which are exposed to X rays. Any discontinuity in the part appears on the film due to differences in X ray absorption.

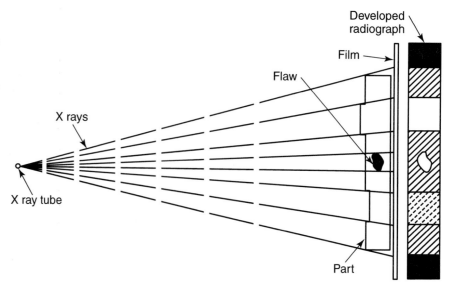

Figure 17-3.
A flaw in the component is shown by a distortion in the magnetic flux, depicted by the elongated magnetic particles.

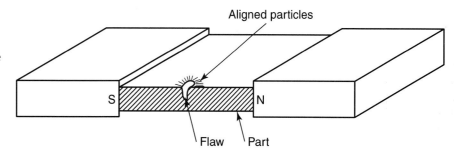

MAGNETIC PARTICLE TEST

Magnetic particle tests can be used only for ferrous alloys. As shown in Figure 17-3, it starts with laying the steel component across the arms of a magnetizing machine. A special metallic powder is sprinkled on the magnetized part surface. Lines of magnetic flux are intensified across a crack. Thus, the magnetic powder is drawn to the crack making it visible.

Magnetic particle tests use both wet and dry magnetic particles to inspect the surface. Wet particles are primarily used for the detection of hairline surface cracks. These particles are typically used in stationary machines where they can be continuously agitated, pumped, and recycled through a spray nozzle. Dry particles are designed for use with portable equipment. They are especially effective on large parts with rough surfaces. The particles themselves are for one-time use and are seldom recovered.

ULTRASONIC TEST

Ultrasonic tests use a "sonar" or sound wave system to detect flaws. Ultrasonic waves are usually produced by the piezoelectric effect using a quartz crystal transducer. The waves are outside the audible range, normally 20 kHz to 50 MHz. The sound wave traveling inside a material is reflected back by any boundary, such as outer surface or an internal flaw.

A transducer sends a wave into the material being tested. The time the sound wave travels in the material depends on the characteristics and thickness of the material. If there is no defect in the material, the sound wave is reflected by the other side of the material. In this case, the total travel time corresponds to the thickness of the material. However, when the sound wave hits any flaw inside the material, the travel time is reduced accordingly. The reflected wave is shown closer to the transmitted beam or initial pulse. Therefore, a flaw is indicated on a screen, as in Figure 17-4. Figure 17-5 shows a typical commercial ultrasound flaw detector. It can be used to detect flaws in all types of materials, including composites.

Figure 17-4.
A—Ultrasonic testing can be used to detect internal flaws. B—The ultrasonic pulse takes a measurable amount of time to travel through a part without flaws. C—When the ultrasonic pulse hits a defect inside the material, the travel time for the pulse is reduced. Therefore, the wave moves closer to the initial pulse.

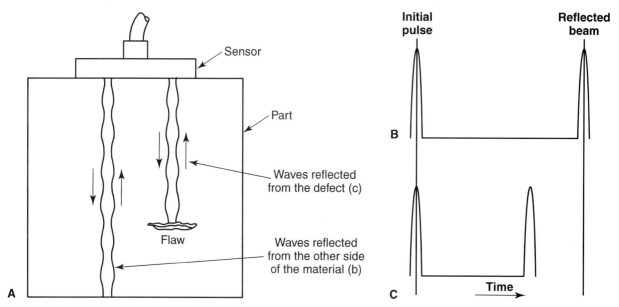

Figure 17-5.
Ultrasonic testing can be used for all materials, including composites.

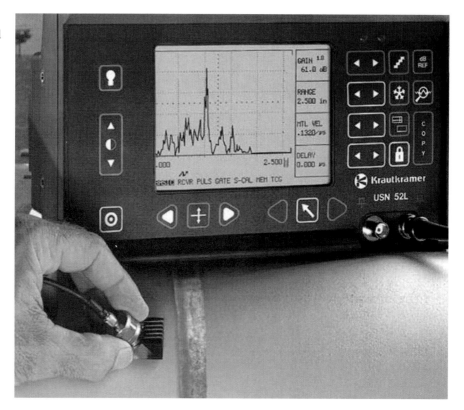

EDDY CURRENT TEST

Eddy current tests involve the use of alternating magnetic fields. These tests can be used on any conducting material. When an alternating current is used to excite a coil, an alternating magnetic field is produced. The magnetic lines of flux are concentrated at the center of the coil. Then, as the coil is brought near an electrically conductive material, the alternating magnetic field penetrates the material and generates continuous, circular eddy currents as shown in Figure 17-6. Larger eddy currents are produced near the test surface. Weaker eddy currents are produced inside the material. Eddy currents induce additional magnetic fields, which interact with the original magnetic field. Any change in the eddy current can be detected by change in the interaction between the original and secondary magnetic fields.

A crack in the test material obstructs the eddy current flow, lengthens the eddy current path, and reduces the secondary magnetic field. This change in secondary magnetic field can be sensed by the primary coil and indicated by appropriate instruments. If a test coil is moved over a crack or defect in a metal part, at a constant clearance and constant speed, a momentary change occurs in coil reactance and coil current. Figure 17-7 illustrates a typical commercial eddy current instrument which can be used to detect cracks, seams, laps, pits, inclusions, and other defects. This instrument can also be used to measure thickness, electric conductivity, magnetic permeability, hardness, or physical dimensions because these properties are all related to the eddy current on the surface.

Figure 17-6.
In an eddy current test, the primary field of the test coil enters test part and generates eddy currents which, in turn, generate a secondary field. Any flaw in the material changes the eddy current and, therefore, can be detected.

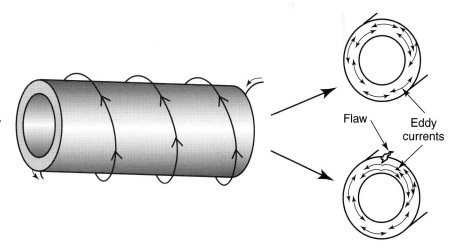

Flaw

Eddy currents

Figure 17-7.
An eddy current instrument
can be used to detect cracks,
seams, laps, pits, inclusions,
and other defects on any
conductive material.
(SE Systems, Inc.)

SUMMARY

Testing of industrial materials is necessary for understanding material behavior. Testing ensures product performance, safety, and quality, and aids in developing new materials. There are two major classes of tests for industrial materials—destructive and nondestructive. Nondestructive tests do not damage or destroy the test part. They are normally used for inspection or detection of defects in materials and structures. Some nondestructive tests are also used for measuring a material's physical and mechanical properties. Most of the nondestructive tests can be conducted quickly and are suitable for automation, which is extremely useful in production line for quality control. For all tests of industrial materials, standard practices are followed. Most of these practices are established by the American Society for Testing and Materials (ASTM), as well as other organizations.

IMPORTANT TERMS

American Society for Testing and Materials (ASTM)

Eddy Current Tests

Magnetic Particle Tests

Nondestructive Test

Penetrant Tests

Radiographic Tests

Ultrasonic Tests

QUESTIONS FOR REVIEW AND DISCUSSION

1. What are the main purposes of testing industrial materials?
2. What is ASTM and what is its main function?
3. Of the six types of standards produced by ASTM, which type would cover tensile strength testing of an aluminum alloy?
4. Of the six types of standards produced by ASTM, which type would cover checking the quality of incoming AISI 4140 steel bars?
5. Which organization governs the standards for concrete materials?
6. What is ANSI and what are the roles of ANSI?
7. What are five common nondestructive tests?
8. Why is a magnetic testing method inappropriate to detect flaws in an aluminum casting?
9. An operator uses an ultrasonic flaw detector to detect flaws inside a steel block. The block is one inch thick. The flaw detector is calibrated so the distance between the initial pulse and the reflected beam is 25. When a flaw is sensed by the detector, the distance is 15. How deep is the flaw from the top surface of the steel block?
10. An ultrasound detector is best used to detect cracks _____ of the part.
11. After a cast iron part is machined, small pin holes are noticed on the machined surface. In an effort to find these defects before the part is machined, a member of the design team suggests using a penetrant test. Is this an appropriate method to detect these flaws?
12. For the application in Question 11, what other methods may be appropriate?
13. Can an ultrasound detector be used to find flaws in polymeric composites?
14. What is the purpose of using ultraviolet light in certain penetrant test methods, such as the inspection of automotive engine blocks?

FURTHER READINGS

1. Davis, H.E., Troxell, G.E. and Hauck, G.F.W. <u>The Testing of Engineering Materials (4th ed)</u>. New York: McGraw-Hill Inc.(1982).

2. Shah, V. <u>Handbook of Plastics Testing Technology (2nd ed.)</u>. New York: John Wiley & Sons (1998).

3. ASTM. <u>2000 Annual Book of ASTM Standards</u>. Philadelphia, PA (2000).

4. Mix, P.E. <u>Introduction to Nondestructive Testing: A Training Guide</u>. New York: John Wiley & Sons (1987).

5. Askeland, D.R. <u>The Science and Engineering of Materials (3rd ed.)</u>. Boston: PWS Publishing Co. (1994).

INTERNET RESOURCES

www.asnt.org
American Society of Nondestructive Testing

www.krautkramer.com
Krautkramer Branson, Inc.

www.astm.org
American Society for Testing and Materials

www.zetec.com
ZETEC

www.geocities.com/SiliconValley/3300
Materials Testing Internet Resources

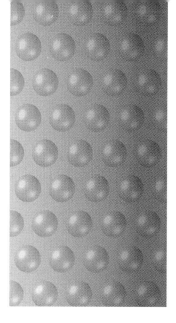

Chapter 18

Introduction to Destructive Testing of Industrial Materials

KEY CONCEPTS

Upon completion of this chapter, you should understand:
➤ How destructive testing differs from nondestructive testing.
➤ Reasons for performing destructive testing.
➤ Common types of destructive tests.
➤ Procedures for conducting destructive tests.

As stated in Chapter 17, the purpose of testing any material is to understand the properties of the material. Properties of materials must be tested for several reasons, including safety and product performance, product quality, new product and process development, and scientific research. This chapter looks at the type of material testing known as destructive testing.

Destructive testing involves partial damage to or complete destruction of the test piece during the testing process. This is different from nondestructive testing. Destructive testing is the most widely used type of material test. It provides information that cannot be obtained through nondestructive testing. Destructive tests include tensile tests, compression tests, shear and torsion tests, impact tests, hardness tests, fatigue tests, and fracture toughness tests.

TENSILE TEST

In a *tensile test,* a gradually increasing uniaxial load is applied to a test piece until the piece fails. In other words, the test piece is pulled apart until it breaks. Properties revealed from a tensile test include ultimate tensile strength, yield strength, modulus of elasticity, percent elongation, and percent reduction in area. These terms are presented and discussed in Chapter 3.

APPARATUS

A testing machine used for a tensile test must be able to apply a tension load and measure the load and elongation of the test piece. The sensitivity and capacity of the machine must conform to ASTM standard E4. A typical tensile test machine is shown in Figure 18-1.

Figure 18-1.
This is a typical tensile
test machine.
(MTS Systems Corporation)

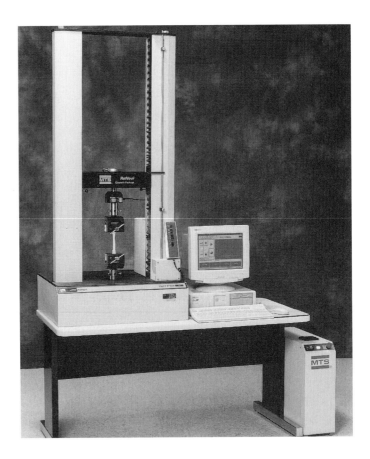

Generally, a load is applied by moving the two cross heads in opposite directions. This movement is controlled in one of the two "loading modes." The first mode is called *loading-rate-control mode.* In this mode, the speed at which the tension force is applied to the test piece is set at a constant level. The second mode is called the *strain-rate-control mode.* In this mode, the rate of test piece elongation is set at a constant level. The "load speed" is adjusted to maintain this level.

The load applied on the test piece is usually measured by a mechanism built into the testing machine. The elongation of the test piece is measured by an extensometer attached to the test piece. With the proper equipment, both load and elongation can be recorded digitally or output on a graph. Stress is calculated as follows.

$$\text{Stress } (\sigma) = \frac{P}{A_o}$$

where P is the load and A_o is the original cross-sectional area. Strain, or elongation, is calculated as follows.

$$\text{Strain } (\varepsilon) = \frac{L - L_o}{L_o}$$

where L is the length of the test piece at a certain stress level and L_o is the original length of the piece.

SPECIMEN

The test piece is called a specimen, or test specimen. A variety of specimen shapes can be used for tension test, such as round, square, or rectangular. The shape used depends on the material being tested. For metals, a round specimen is commonly used. For sheet, plate, or tube materials, a flat specimen is generally used. Figure 18-2 illustrates a typical specimen design. The central region normally has a reduced cross section. Both ends are machined suitable for firm gripping. The ends of a round specimen can be plain, shouldered, or threaded. Rectangular specimens are generally made with plain ends, but sometimes they are shouldered or have a hole for a pin bearing. For detailed specimen dimensions, refer to ASTM Standard E8 for metallic materials and D638 for plastics.

The tensile strength of concrete can be measured using *splitting tension test.* This test is governed by ASTM Standard C496. As shown in Figure 18-3, the splitting tension test uses a standard 6″ × 12″ (150 mm × 300 mm) cylinder. The cylinder is loaded in compression along the diameter. Narrow strips of plywood can be used as a cushioning material along the loading line. The splitting tensile strength is calculated from:

$$\sigma = \frac{2P}{\pi l d}$$

where σ is the splitting tensile strength, P is the maximum applied load, l is the length, and d is the diameter. This type of test is much more simple than an axial tension test and is the only ASTM standardized tension test for concrete.

COMPRESSION TEST

A *compression test* is the opposite of a tension test, with respect to loading direction. Instead of pulling apart a specimen, it is pressed together. Compression tests are generally used for brittle materials in applications

Figure 18-2.
A tensile test specimen usually has a reduced section in the central test region with both ends suitable for gripping.

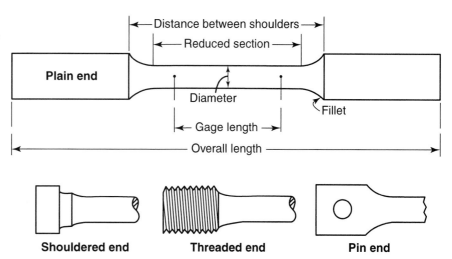

Shouldered end   Threaded end   Pin end

Figure 18-3.
Splitting tension test uses
compression load to test the
tensile strength of concrete.

where resistance to compressive forces is necessary, such as building foundations and machine bases. Examples of brittle materials include cast iron, mortar, concrete, brick, and ceramic products. Tensile strengths of these materials
are usually low compared with their compression strengths.

APPARATUS

Universal testing machines can be used for compression testing, as
shown in Figure 18-4. These machines are generally used for laboratory
testing. There is also field equipment designed specifically for compression

Figure 18-4.
Universal testing machines
can be used for compression
testing in the laboratory.

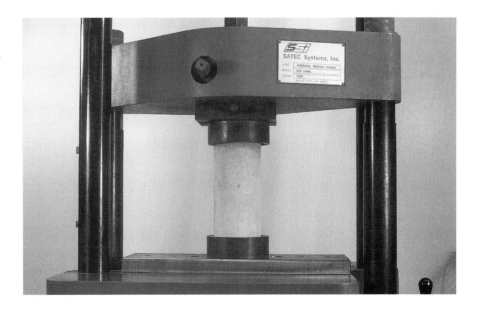

testing. Many are used for routine compression tests of concrete specimens in the field or laboratory. Normally, these machines are less expensive than universal testing machines. Since they are portable, they are also suited for mobile quality control laboratories at a construction site.

SPECIMEN

Standard laboratory test specimens for concrete are cylindrical with a height-to-diameter ratio of two. They are cast and hardened in an upright position. The standard specimen is normally a 6″ × 12″ (150 mm × 300 mm) cylinder where the maximum size of coarse aggregates does not exceed 2″ (50 mm). If the maximum size of coarse aggregates exceeds 2″ (50 mm), the concrete sample must be treated by wet sieving as described in ASTM standard C172. An alternative is to increase the diameter of the cylinder to at least three times the nominal maximum size of coarse aggregate in the concrete. Refer to ASTM standard C31 for more details. Specific ASTM standards should be consulted for other materials or products, such as:

- Metallic materials: ASTM E9
- Wood: ASTM D143, D198
- Drain tile: ASTM C4
- Structural clay tile: ASTM C67
- Sewer pipe: ASTM C14
- Refractory brick: ASTM C133
- Vulcanized rubber: ASTM D395, D575
- Building stone: ASTM C170

Both ends of a compression specimen should be flat to prevent stress concentrations. The end surfaces should also be perpendicular to the cylinder axis to prevent bending due to eccentric loading. Moreover, most brittle materials are sensitive to the loading rate. The rate should be low for compression testing. The recommended load rate for concrete is between 1,000 lb/min and 2,000 lb/min.

SHEAR AND TORSION TESTS

Applications such as rivets, crank pins, and wooden blocks are subject to shearing forces. Therefore, shear strength must be tested to ensure the safety and functionality of the part. A specific kind of shear stress occurs inside a shaft subject to twisting. This is called *torsion.* Therefore, a torsion test is used to evaluate the shear properties of materials as well.

DIRECT SHEAR

For a direct shear test of metals, a bar is usually sheared in a device that clamps part of the specimen while the remaining portion is subjected to load. Any means of applying a load, such as a hydraulic system or a universal test machine, can be used. Shear strength is defined as:

$$\text{Shear Strength } (\tau) = \frac{P}{A_o}$$

where P is the maximum force required to shear the specimen and A_o is the area of the shear plane parallel to the loading direction. The shear plane can be either rectangular or circular. For direct shear tests of metals, refer to ASTM standard B565.

In Johnson shear test, a rectangular bar about $1'' \times 2''$ (25 mm $\times$ 50 mm), or a cylindrical rod about $1''$ (25 mm) diameter is used. As shown in Figure 18-5, specimen A is clamped to the base C. A force applied to the loading tool E ruptures the specimen in single shear. If the specimen is extended to B and bridges the gap between the dies D, it is subjected to double shear. The dies and the loading tool are made of tempered tool steel ground to an edge.

For direct shear tests of wood, a special tool and specimen are used, as shown in Figure 18-6. Refer to ASTM D143 for specific information. Failure tends to occur along the shear plane marked by the dashed line shown in Figure 18-6. For shear tests of glued joints, a specimen is glued along the dashed line (shear plane) and then loaded.

TORSION

A torsion test of metals is usually done in a specially designed torsion testing machine. An illustration of one type of torsion testing machine is shown in Figure 18-7. A driving mechanism turns a chuck that clamps onto

Figure 18-5.
For a shear test, a bar or a plate is sheared in a device which clamps part of the specimen while the remaining portion is subjected to load.

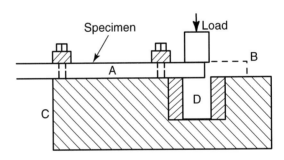

Figure 18-6.
This schematic shows the method of testing wood in direct shear.

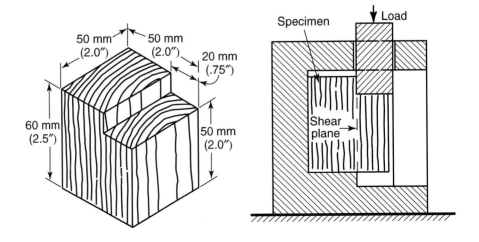

one end of the specimen. The chuck turns and applies torque through the test specimen to a similar chuck at the other end. This chuck measures the torque.

Before a test is started, the specimen is carefully measured. In the elastic range, maximum fiber stress τ is related to the torque T by the torsion formula for circular shafts:

$$\tau = \frac{Tr}{J}$$

where r is the outside radius and J is the *polar moment of inertia* of the cross section. The polar moment of inertia J for a solid circular bar is:

$$J = \frac{\pi d^4}{32}$$

For more details of the test procedure, refer to ASTM standard E558.

IMPACT TEST

The tests discussed to this point basically measure the static properties of materials. However, many structures and components are subjected to dynamic loading. An important type of dynamic loading occurs when a load is applied suddenly, such as the impact of a moving object. For applications where dynamic load is anticipated, an impact test is necessary to determine material toughness. Examples of these applications include parts in percussion drilling equipment, automobile engines, transmissions, body structures, tracks, buffer devices, and highway guardrails.

Figure 18-7.
In a torsion test, a chuck applies torque through the test specimen. An indicator displays the torque. (Tinius Olsen Testing Machine Co.)

Specimen

The velocity of an object is reduced as it strikes another object. Part or all of the first object's kinetic energy is transferred to the second object. A material that absorbs larger amounts of energy before fracturing is said to have high toughness. A material with high toughness also has better resistance to impact loading.

An *impact test* is a dynamic test where a specimen is struck and broken by a single blow. The test is done in a specially designed testing machine, Figure 18-8. The energy absorbed in breaking the specimen is recorded. This is done by measuring the difference between the pendulum heights before and after the specimen is broken. The applied load may be flexure, tension, compression, or torsion. Flexural loading is the most common. Tensile loading is less common. Compressive and torsional loads are used only in special cases.

The most commonly used impact tests for steels are Charpy and Izod. Both tests use the pendulum principle. These tests are conducted on small, notched specimens as shown in Figure 18-9. In a Charpy test, the specimen is supported as a simple beam. In an Izod test, the specimen is supported as a cantilever. Procedures for the Charpy and Izod testing of metals is standardized by ASTM E23. Formal material specifications of impact-strength limits is established for a number of products, such as airplane engine parts, transmission gears, parts for tractor belts, turbine blades, forgings, and pipes. Charpy and Izod tests are also used for plastics and electrical insulating materials. Refer to ASTM D256 for specifics on testing these materials.

Figure 18-8.
In an impact test, a pendulum is released at prescribed height. The energy absorbed by the specimen is measured by the difference in heights before and after the specimen is broken.
(Tinius Olsen Testing Machine Co.)

Figure 18-9.
A—In the Charpy test, the specimen is supported as a simple beam. B—In the Izod test, the specimen is supported as a cantilever.

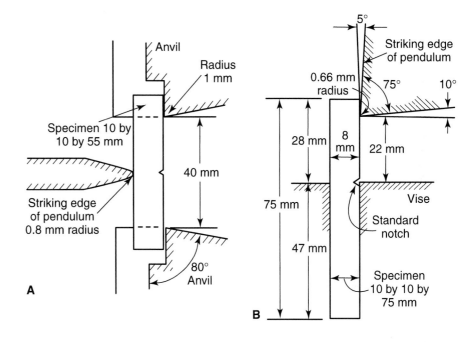

HARDNESS

Hardness is defined as the resistance of a material to surface deformation, particularly permanent deformation, indentation, or scratching. Hardness tests are widely used since they are quick and easy to perform. Similar materials can be graded according to their hardness. In addition, the quality of materials can be checked and controlled using hardness because of the relationship between hardness and other mechanical properties. For example, having established a correlation between hardness and tensile strength, a simple hardness test can then be used to predict and control the tensile strength of the product.

There are many types of hardness tests. This section discusses three typical hardness tests. These are the Brinell test, Rockwell test, and Durometer test.

BRINELL HARDNESS TEST

Brinell hardness test consists of pressing a hardened steel ball into a specimen. In accordance with ASTM standard E10, a 10 mm diameter ball is typically used. A load is exerted by a mass of 3,000 kg for hard materials, 1,500 kg for materials of intermediate hardness, and 100 kg to 500 kg for soft materials. A typical Brinell hardness tester is illustrated in Figure 18-10.

Normally, the full load is applied for a minimum of 15 seconds for ferrous metals and 30 seconds for softer metals. Then, the load is released and the diameter of the indentation is measured to the nearest 0.02 mm using a microscope. The *Brinell hardness number (BHN)* is defined as the amount of load per unit area (kg/mm^2) based on the area of the indentation that remains after the load is removed. The BHN is calculated by dividing the applied load by the area of the indentation surface, which is assumed to be spherical. If P is the applied load, D is the diameter of the steel ball, and d is the diameter of the indentation, then

$$\text{Brinell hardness number (BHN)} = \frac{P}{\frac{\pi D}{2}\left(D - \sqrt{D^2 - d^2}\right)}$$

A Brinell hardness number can also be calculated from the depth of indentation using the following equation:

$$\text{Brinell hardness number (BHN)} = \frac{P}{\pi D t}$$

where P is the applied load, t is the depth of indentation, and D is the diameter of the ball.

Figure 18-10.
A digital Brinell hardness tester. (Wilson Instruments)

ROCKWELL HARDNESS TEST

The ***Rockwell hardness test*** is similar to the Brinell test. The hardness number is based on the depth of indentation an indenter makes on the test piece. Different loads and indenters are used for different Rockwell hardness scales, depending on test conditions. Rockwell hardness tests are widely used in industrial work. The procedures are standardized by ASTM E18. Figure 18-11 shows a digital Rockwell hardness tester.

Figure 18-11.
A digital Rockwell hardness tester. (Wilson Instruments)

Two common Rockwell hardness scales are Rockwell B (HRB) and Rockwell C (HRC). These are calculated:

$$\text{Rockwell B (HRB)} = 130 - \frac{\text{depth of penetration (} \mu m)}{2 \ \mu m}$$

$$\text{Rockwell C (HRC)} = 100 - \frac{\text{depth of penetration (} \mu m)}{2 \ \mu m}$$

A Rockwell scale is divided into 100 divisions. Each division, or "point" of hardness, is the equivalent of 2 μm indentation. The higher the Rockwell hardness number, the harder the material. In an actual test, the above calculation is automatically done by the tester.

Various Rockwell scales are listed in Figure 18-12. In general, select a scale so that the expected results fall in the range. For example, the working range for B scale is from HRB 0 to HRB 100. If the material is harder than HRB 100 or the material is softer than HRB 0, the hardness test is invalid. The range for Rockwell C scale is HRC 20 to HRC 70. This scale is most commonly used for materials harder than HRB 100.

Figure 18-12.
This chart shows typical Rockwell hardness scales.

ROCKWELL HARDNESS SCALES			
Group	**Scale Symbol**	**Indenter**	**Major Load (kg)**
1 (common scales)	B	1/16″ ball (1.6 mm)	100
	C	Diamond cone	150
2	A	Diamond cone	60
	D	Diamond cone	100
	E	1/8″ ball (3.2 mm)	100
	F	1/16″ ball (1.6 mm)	60
	G		150
	H	1/8″ ball (3.2 mm)	60
	K		150
3	L	1/4″ ball (6.4 mm)	60
	M		100
	P		150
	R	1/2″ ball (12.7 mm)	60
	S		100
	V		150

DUROMETER HARDNESS TEST

The ***Durometer hardness test*** is mostly used for measuring the relative hardness of soft materials, such as plastics and rubber. This test method is based on the penetration of an indenter forced into the material under specified conditions. Figure 18-13 shows an automatic Durometer system used to test plastic and rubber.

Durometer hardness tester consists of a pressure foot, an indenter, and a display and/or a data port to a computer. A test is carried out by first placing a specimen on a hard, flat surface. The pressure foot is pressed onto the specimen, making sure that it is parallel to the surface of the specimen. The indenter is forced into the specimen and the hardness is calculated by the tester.

Figure 18-13.
In this precision Durometer hardness test system, the indenter penetration is electronically measured, and converted to a Durometer hardness number.
(Instron Corporation)

Two types of Durometers are most commonly used—Type A and Type D. The difference between the two types is the shape and dimension of the indenter. The Type A Durometer test is used with relatively soft material. The Type D Durometer test is used with slightly harder material. The testing method is governed by ASTM standard D2240.

FATIGUE TEST

Most structures are subject to varied loads which causes fluctuations in applied stresses. With dynamic loading, though the maximum stress may be considerably less than the static strength of the material, failure may occur when the stress is repeated often enough. A failure induced in this manner is called a *fatigue failure.* The stress at which a material fails by fatigue after a certain number of cycles is termed the *fatigue strength.*

For most materials, there is a stress limit below which a load can be repeatedly applied without causing failure. This limiting stress is known as the *endurance limit* or *fatigue limit.* The magnitude of the fatigue limit depends on the kind of stress variation the material is subjected to. The purpose of most fatigue tests is to find out the fatigue strength and endurance limit of a material.

There are several fatigue testing devices. These are typically simple, yet accurate, devices. A typical tension fatigue tester is shown in Figure 18-14. This tester can apply loading cycles in a high frequency. Another commonly used fatigue tester is a rotating beam type, shown in Figure 18-15. A specimen is held by its ends and loaded through two bearings equidistant from the center of the span. Two equal loads are applied on these bearings, producing a uniform bending of the specimen. The top of the specimen is subjected to maximum compressive stress. The bottom of the specimen is subjected to maximum tensile stress. Therefore, when the specimen is rotated, a given point on the specimen's radius is alternating from maximum compressive stress to maximum tensile stress. Therefore, a complete stress cycle is produced in the specimen.

The specimen is subjected to pure bending. The bending moment is determined by the following formula.

$$M = \frac{W}{2} L$$

where W is the load and L is the span between the two bearings. The maximum tensile or compressive stress (σ) on the specimen due to bending is:

$$\sigma = \frac{Mr}{I}$$

where r is the radius and I is the moment of inertia of the specimen. For a circular specimen, I is given by the following equation.

$$I = \frac{\pi r^4}{4}$$

To determine the fatigue limit of a material, it is necessary to prepare a supply of identical specimens representative of the material. The first specimen is tested at a relatively high stress, so failure occurs after a small number of stress cycles. Other specimens are then tested at lower stress levels. The number of repetitions required to produce failure increases as the stress decreases. Specimens stressed below the fatigue limit will not fail.

A stress (S) versus number of cycles (N) curve can be produced. See Chapter 3. The fatigue strength and endurance limit can be determined from this curve. Refer to ASTM E606 for fatigue tests of metals and ASTM D671 for fatigue tests of plastics.

Figure 18-14.
A typical tension fatigue tester.
(MTS Systems Corporation)

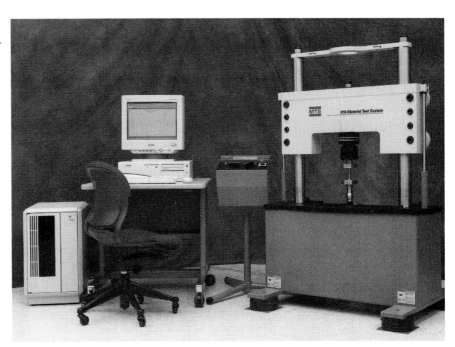

Figure 18-15.
A rotating beam fatigue-testing machine generates a complete reversed loading cycle per revolution.

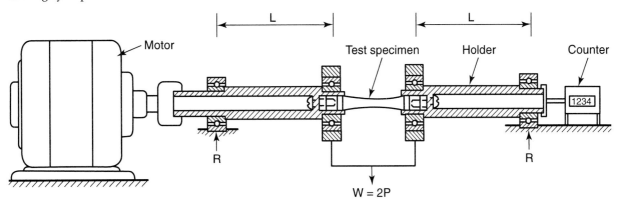

FRACTURE TOUGHNESS TEST

A *fracture toughness test* measures the resistance of a material to fracture. In other words, it measures a material's ability to stop a crack from spreading. For a thick material with a small internal crack, the fracture toughness (K_{IC}) is defined as:

$$K_{IC} = \sigma_c \sqrt{\pi a}$$

where σ_c is the critical stress in the material as the cracking is started and "a" is the pre-existing crack length.

SPECIMEN

Figure 18-16 illustrates a typical three-point bend specimen and its loading configuration used to measure fracture toughness of metals. There are two important considerations for specimen preparation. The first consideration is on the dimension requirements of the specimen. To ensure that results are not affected by the test specimen geometry, minimum requirements on thickness (B) and precrack length (a) are as follows.

$$a \geq 2.5 \left(\frac{K_{IC}}{\sigma_Y}\right)^2$$

$$B \geq 2.5 \left(\frac{K_{IC}}{\sigma_Y}\right)^2$$

where σ_Y is the yield strength of the material to be tested. However, in practice K_{IC} is not known before testing. Thus, the value has to be estimated. If the value which results from the test does not meet the requirements, the test must be invalidated for predicting K_{IC} and a new test must be run on a larger specimen.

Another important consideration for test specimen preparation is the crack processing. Normally, a notch is machined into the specimen. Then, a fatigue crack is allowed to develop by applying controlled loading cycles and stress. Thus, a specimen with a sharp crack perpendicular to the specimen sides can be made. The ratio of precrack length (a) to the height of the specimen (W) should be between 0.4 and 0.6.

Figure 18-16.
A test specimen and loading configuration of a three-point bend test for fracture toughness (K_{IC}).

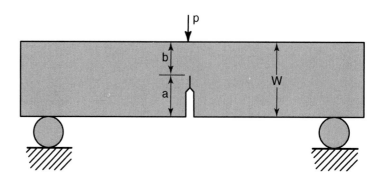

PROCEDURE

A three-point bend specimen can normally be used with a universal test machine. The load (P) is applied to the specimen until the specimen fractures. After the maximum load to fracture the specimen is recorded, the stress intensity factor, or fracture toughness (K_{IC}), can be calculated according to the following formula.

$$K_{IC} = 3.75 \frac{PW}{B(Wa)^{3/2}}$$

where P is the maximum load to break the specimen, W is the height of the specimen, B the thickness of the specimen, and "a" is the length of the precrack.

SUMMARY

Testing of industrial materials is necessary for understanding material behavior; ensuring product performance, safety, and quality; and assisting in development of new materials. There are two major classes of tests for industrial materials—destructive testing and nondestructive testing. A destructive test is used mostly for understanding the strength, ductility, impact resistance, and fracture behavior of materials. This type of test induces permanent damage or destruction on the test piece. For all tests of industrial materials, standard practices are followed. Most of these practices are from the American Society for Testing and Materials (ASTM), as well as other organizations.

IMPORTANT TERMS

Brinell Hardness Number (BHN)
Brinell Hardness Test
Compression Test
Destructive Testing
Durometer Hardness Test
Endurance Limit
Fatigue Failure
Fatigue Limit
Fatigue Strength
Fracture Toughness Test

Hardness
Impact Test
Loading-Rate-Control Mode
Polar Moment of Inertia
Rockwell Hardness Test
Splitting Tension Test
Strain-Rate-Control Mode
Tensile Test
Torsion

QUESTIONS FOR REVIEW AND DISCUSSION

1. How is the tensile strength of a concrete mixture determined? Which ASTM standard covers this procedure?
2. Why does the compression strength of building stone have to be tested? What ASTM standard will be used for the testing?
3. What is torsion?

4. What is a direct shear test?

5. What does an impact test measure?

6. What situations will the impact test be applied to?

7. What does a hardness test measure?

8. For a standard Brinell hardness test, a 10 mm ball and a load of 3,000 kg are used. After the load is applied and then released, the indentation is measured using a microscope to be 2.737 mm in diameter, what is the Brinell hardness number (BHN)?

9. If a Rockwell hardness test has been performed on a steel and the hardness is determined to be HRC 40, what is the depth of indentation on the steel surface caused by the test?

10. Two steel samples are Rockwell hardness tested. One has a hardness of HRC 50 and the other HRC 30. Which steel is harder?

11. What type of materials is the Durometer hardness test used for?

12. In a fracture toughness test, a maximum stress of 10,000 psi was measured in order to break a three-point bend specimen with a precrack of 0.25″. What is the fracture toughness for this material?

FURTHER READINGS

1. Davis, H.E., Troxell, G.E. and Hauck, G.F.W. The Testing of Engineering Materials (4th ed.). New York: McGraw-Hill Inc. (1982).

2. Shah, V. Handbook of Plastics Testing Technology (2nd ed.), New York: John Wiley & Sons (1998).

3. ASTM. 2000 Annual Book of ASTM Standards. Philadelphia, PA (2000).

4. Askeland, D.R. The Science and Engineering of Materials (3rd ed.). Boston: PWS Publishing Co. (1994).

INTERNET RESOURCES

www.instron.com/index1.html
Instron Corp.

www.mts.com
MTS Systems Corp.

www.astm.org
American Society for Testing and Materials

www.geocities.com/SiliconValley/3300
Materials Testing Internet Resources

Glossary

A

δ Iron: Pure iron above 2541°F (1394°C).

α Iron: Pure iron below 1674°F (912°C).

γ Iron: Pure iron between 1674°F (912°C) and 2541°F (1394°C).

Ablation: Physical property of wood related to its ability to char when exposed to fire, but retain most of its strength.

Absorption: Refers to the property of a material absorbing a portion of light.

Accelerators: Chemical added to expedite the curing process.

Acceptor Impurities: Chemical elements from Group III of the Periodic Table of Elements used as dopants in semiconductors to affect the flow of electrons.

Acrylic: Common name of polymethyl methacrylate (PMMA).

Addition Polymerization: A type of polymerization often referred to as *chain-growth polymerization* in which heat and pressure are used in combination with an initiator to break a double-bonded monomer, causing a polymer to form.

Advanced (Engineered) Ceramics: A classification used to define a group of ceramics tailored or engineered to meet specific operational conditions, usually for high-technology applications.

Age Hardening: A process where, after solution treatment, an aluminum alloy is held at a low temperature for certain period of time, which strengthens and hardens the aluminum alloy.

Aggregate: Any of several hard inert materials (sand, gravel) used for mixing with a cement to form concrete.

Air Entraining: A process of adding a chemical, sodium lauryl sulfate, to Portland cement causes the formation of tiny bubbles in the concrete allowing it to expand and contract without failing.

Allotrope: A type of polymorph that can change its structure with changes in its environment.

Alloy Steels: The total percentage of alloying elements in the steel is less than or equal to approximately 5 percent.

Alloying Elements: Elements added into iron and steel to improve properties.

Alumina Silicates: Ceramic compounds formed from oxygen, silicon, and aluminum.

Aluminum Bronze: A no-tin alloy with up to 10% aluminum.

Amorphous Polymer: Polymer with its molecules randomly distributed.

Anion: Negatively charged ion, more electrons than protons.

Anisotropic: Exhibiting properties with different values when measured along axes in different directions.

Annealing: A process of heating a steel or cast iron to its austenitizing temperature and then cooling it slowly normally in a furnace.

Annual Rings: A band of earlywood and latewood cells that form with each year's growth.

Anodizing: A process in which a thick oxide coating is formed on the metal surface with the help of an electrochemical reaction.

Aramid Fibers: Fibers made from highly aromatic polymers. Aramid stands for "aromatic polyamide."

Aspect Ratio: The physical length-to-diameter ratio of the reinforcement of a composite material.

ASTM: American Society for Testing and Materials.

Atom: Smallest part of an element that can retain the physical and chemical properties of that element.

Atomic Mass: Average mass of an element.

Atomic Mass Unit (u): Unit designating the atomic mass of an element. An atomic mass unit is defined as 1/12th the mass of carbon 12.

Atomic Number: The number of protons in a given chemical element.

Atomic Weight: Average weight of all isotopes in an element.

Austempering: A process of quenching a hardenable steel or cast iron to a temperature, holding the part at this temperature until transformation is completed, and then air-cooling the part.

Austenite: A face centered cubic γ iron with carbon dissolved in between iron atoms.

Austenitic Stainless Steels: Stainless steel having austenitic (FCC, γ-iron type) structure at all normal heat treatment temperatures.

Austenitizing: A process of heating steels or cast irons to high temperature so the steel or cast iron is transformed into fully austenite.

Autoclave Curing and Bonding: A process in which an entire composite assembly is placed in a heated and pressurized chamber for accelerated curing.

Avogadro's Number: The number 6.023×10^{23}, which is the number of atoms of any substance in one mole of the element.

B

Baghouses: Filters used in the manufacture of Portland cement which remove particles from exit gases before they enter the atmosphere.

Bainite: A microstructure resulted from austempering, which has a similar mechanical property to tempered martensite.

Ball Clay: Sedimentary clay of excellent plasticity, used extensively in manufacturing whiteware.

Band of Energy Levels: Sublevels of the electron shells that orbit the nucleus of an atom.

Bending Strength: A measure of the stiffness of a material, or the ability of a material to resist a force or load applied.

Beryllium Copper: A copper-beryllium alloy.

Branched Polymer: A polymer of molecules that has some branches, like a tree branch.

Brass: A copper-zinc alloy.

Brinell Hardness Number: A measure of hardness defined as the amount of load per unit area, in kg/mm^2, based on the area of the indentation that remains after the load is removed.

Brinell Hardness Test: A hardness test based upon indentation area.

Brittle Fracture: A fracture which occurs after a material undergoes very limited plastic deformation.

Bronze: Refers to any copper alloy except a copper-zinc alloy (brass).

Buckling: The action of a material bending or deflecting laterally when subjected to compressive force.

C

Cambium Layer: A thin layer of cells outside of the sapwood which is the newest growth.

Carbon Fiber: Reinforcing fiber made of carbon, normally heat treated at a lower temperature during final stage.

Carbon/Graphite (CG) Fibers: Organic precursor fibers, such as polyacrylonitrile (PAN), used as reinforcement in composites.

Carburizing: A process in which carbon is introduced into the steel surface layer at high temperature.

Cartridge Brass: A copper alloy containing 30% zinc, which can be used for deep drawing.

Cast Iron: Ferrous alloy containing 2 percent to 4 percent of carbon.

Catastrophic Oxidation: The accelerated oxidation attack.

Cation: Positively charged ion, more protons than electrons.

Cellulose $(C_6H_{10}O_5)$: A naturally occurring thermoplastic polymer.

Cement: A finely pulverized material consisting of compounds of lime, silica, alumina, and iron.

Cementite: A compound of iron and carbon (Fe_3C).

Ceramic Matrix: A binder in a ceramic composite that is itself classified as a ceramic, usually a powder.

Ceramic Matrix Composite (CMC): One or more ceramic materials or ceramic material phases combined to provide a material with superior properties to those of its individual constituents.

Ceramics: Hard brittle compounds of metallic and nonmetallic elements that have high melting temperatures and are chemically inert.

Charge: A stack of sheet molding compound plies placed in a mold.

Chemical Properties: Measures of how a material interacts with an environment.

Chemical Vapor Deposition (CVD) Infiltration: A continuous process in which infiltration is performed after CVD coating of fibers.

Chemical Vapor Infiltration (CVI): A process used to agglomerate ceramic powders by introducing a gas into the composite at the reinforcing stage. The gas decomposes then decomposes to form a solid matrix.

Clinker: Marble-size pieces of the raw materials of Portland cement prior to being ground into a fine powder.

Coefficient of Linear Thermal Expansion: A material constant which represent the length change due to thermal expansion per unit length and per degree of temperature increase.

Coercive Field: Also known as coercive magnetic force.

Coercive Magnetic Force: A magnetic force (strength) needed in order to lower the flux density to zero.

Coke: Fuel made from coal.

Composite: One or more materials or material phases combined to produce a material that has superior properties to those of its constituents.

Compression Molding: A forming process used with plastics which employs heat and pressure directly on the resin in a closed mold.

Compression Strength: The ability of a material to withstand a pressing or squeezing stress.

Compression Test: A test procedure of applying compressive load on a specimen until failure.

Concrete: A monolithic ceramic product of aggregates bonded with cement, such as Portland or asphalt.

Condensation Polymerization: A type of polymerization in which a catalyst causes a chemical reaction to occur which results in a polymer forming from a monomer with a condensate, e.g. water, as a by-product.

Conduction Band: Area in which electrons that have been freed from chemical bonds reside at a much higher energy level.

Conductivity: The reciprocal of resistivity for any material.

Coniferous: Cone bearing species of trees that retain their leaves; source of softwood lumber.

Copolymer: A polymer formed when two or more types of monomers are included in the polymer molecule.

Corrosion: The destruction or deterioration of a material because of reaction with its environment, usually in the forms of materials removal and oxide formation.

Corrosion Fatigue: A form of corrosion that is accelerated by a cyclic or repeated stress.

Coupling Agents: Materials, usually liquids, that promote the interface adhesion between the matrix and the reinforcement in a composite material.

Covalent Bond: Atom-to-atom chemical bond in which the valence shell of each atom shares its electrons to achieve the lowest energy level.

Crack Propagation: A property usually associated with ceramics and concrete. When ceramic or concrete products crack, the crack often grows (propagates) due to the brittle nature of these materials.

Creep: Deformation continues to increase with time when a constant load is applied on the material.

Crystalline Ceramics: Ceramic bodies having an orderly arrangement of atoms which form lattice structures of repeating patterns.

Crystalline Solid: Solids with ordered three-dimensional geometric patterns that form their structure.

Curing: A process during which a thermoset plastic hardens.

Cyclic Loading: A mechanical force that changes repeatedly during service.

D

Damage Tolerance: Ability of a structure to function adequately after a flaw has been introduced.

Deciduous: Broadleaf species of trees that lose their leaves each year; source of hardwood lumber.

Densification: Three-step process used to reduce porosity of ceramics.

Density: Mass or weight per unit volume.

Deoxidized Copper: Copper with phosphorus added to remove oxygen.

Destructive Testing: A material test which results in damage or destruction of test specimen.

Diamond Grinding: A material removal process that uses a diamond wheel to shape and finish ceramic composites.

Die Pressing: A forming process using fine ceramic powders that are forced together under tremendous pressure then heated causing them to stick together.

Dielectric: A nonconducting material.

Dielectric Constant: A factor by which capacitance will be increased relative to vacuum if a dielectric material replaces vacuum in a capacitor.

Diffuse Reflection: Light reflection from a rough surface.

Diffusion Bonding: A process of hot-pressing an array of fibers between metal foils to form composites.

Donor Atoms: Atoms of chemical elements found in Group V of the Periodic Table of Elements that donate electrons to the conduction band.

Dopants: Chemical elements added to semiconductor materials to affect the flow of electrons.

Dry Process: One of two processes used to manufacture Portland cement. During this process the raw materials are ground dry, properly proportioned, and fed into a kiln in a dry state.

Ductile: A material that tends to stretch or deform appreciably before fracture.

Ductile Cast Iron: Iron having spherical graphite, also known as nodular cast iron.

Ductile Fracture: A fracture which occurs after a material undergoes a large amount of plastic deformation.

Durometer Hardness Test: A hardness test for polymers based on needle penetration.

E

Early Wood (Spring Wood): One of two bands of cells that form the annual rings on trees and which represent growth occurring during the spring of the year.

Eddy Current Test: A nondestructive test utilizing eddy current on the tested part due to magnetic field.

E-glass: A type of glass fiber used in the manufacture of fiberglass parts and products. The E stands for electrical because it has excellent dielectric properties.

Elastic: The property of a material returning to its original shape after removing a load.

Electrical Conductivity: Ability of a material to conduct electrons.

Electrical Discharge Machining (EDM): A material removal process that uses a controlled arcing between the electrode (cathode) and the workpiece (anode).

Electrolytic Tough Pitch (ETP) Copper: A copper produced using an electrolytic tough pitch (ETP) process. It has a minimum of 99.9% copper and a nominal 0.04% oxygen.

Electron Cloud: Term used to refer to the accumulation of electrons in metals that forms during chemical bonding.

Emission: Electromagnetic radiation from a material.

Endurance Curve: A curve plotting stress versus the number of cycles to failure.

Endurance Limit: A stress at which a material will not fail due to fatigue.

Energy Levels: Orbits surrounding the nucleus of an atom that contain the electrons.

Environmental Stress Cracking: A failure that occurs with a specific combination of material and environmental media.

Epoxies: Polymers synthesized from resins characterized as having in their molecules a highly reactive oxirane ring of triangular configuration.

Equilibrium Moisture Content (EMC): A condition where the moisture content in the air and moisture content in wood are equal.

Extensometer: A sensor measuring the elongation or strain of a sample.

Extrinsic Semiconductors: Semiconductors with carefully controlled amounts of impurities or dopants.

Extrusion: Method of forming products by forcing material through a die whose cross section is in the shape of the finished product.

Extrusion Blow Molding: A blow molding technique.

F

Factory and Shop Lumber: A classification of softwood lumber used primarily for structural purposes.

Fatigue: The reducing of a stress limit due to cyclic loading.

Fatigue Failure: A type of mechanical failure occurring when the maximum stress is considerably less than the static strength of the material (as tested in a tension test) under dynamic loading conditions.

Fatigue Limit: A stress level below which a material will not fail due to fatigue.

Fatigue Strength: A stress at which a material fails by fatigue after a certain number of cycles.

Ferrite: A body centered cubic α iron with carbon dissolved in between iron atoms.

Ferrous Alloys: Iron-based alloys that consist of certain amounts of carbon, normally referred to as iron and steel.

Ferrous Metals: Metals having significant iron content and are magnetic.

Fiber: A type of reinforcement with a high aspect ratio.

Fiber Reinforced Superalloy (FRS): Oxidation-resistant high alloy reinforced by fibers.

Fiber Saturation Point (FSP): A point during the seasoning (drying) process when all water in the cell cavities of the wood has evaporated, but the fibers of the cell are fully saturated.

Fiberglass: A composite of glass fiber embedded in a matrix (adhesive) of polymer resin.

Filament Winding: A composite manufacturing process in which a continuous tape of resin-impregnated fibers is wrapped over a mandrel to form a part.

Filler: An additive to polymers with an aspect ratio close to one.

Fireclay: High-melting-temperature clays commonly used for furnace linings, flues, and firebrick.

Flake: A reinforcement material classified as a particulate. Particulates have low aspect ratios. Flakes are thin and short.

Flame Hardening: A surface hardening process using a torch to rapidly heat the surface of a steel workpiece into the austenitic condition which is then quickly quenched.

Flexural Strength: Ability of a material to resist a bending force or load applied perpendicular to its horizontal axis.

Flint Clay: A hard, low plasticity clay used for mixing in ceramics to reduce drying and firing shrinkage.

Fluorescence: A glow of light produced by a material when an ultraviolet stimulus is applied.

Fracture Toughness: The ability to resist fracture.

Friable: Literally means "crumbles;" refers to the ability of a grinding wheel to release pieces, of the abrasive when in contact with a brittle metal.

Frit: Glass powders used in making porcelain enamels.

G

Galvanic Corrosion: A form of corrosion that occurs when two or more dissimilar metals are in contact, one of which corrodes much faster.

Galvanized Steel: A steel coated with zinc for corrosion resistance.

Gas Carburizing: A process in which carbon is introduced to the surface layer of a steel through gaseous chemical reaction.

General Corrosion: Also known as uniform attack (corrosion).

Glass Transition: A change from rubbery state to glassy state for polymers.

Glass Transition Temperature: A temperature for a polymer below which the polymer is in a glassy state and above which it is rubbery.

Glazing: Applying a glassy coating to seal the surface of ceramic products, especially whitewares.

Graphite Fiber: Reinforcing fiber made of carbon, normally heat treated at a high temperature during final stage.

Gray Cast Iron: Cast iron with graphite flakes.

Green Forming: Shaping ceramic bodies while the clay compounds have sufficient moisture to allow for plastic deformation.

Green Lumber: Newly cut lumber with high moisture content.

Groups: Term used to identify the columns in the periodic table of elements.

H

Hard Magnet: Material that can be magnetized easily and stays magnetic when the power is off.

Hardenability: Represents the ease of a steel or cast iron to obtain hardening martensite.

Hardness: Resistance of a material to surface deformation, particularly permanent deformation, indentation, or scratching.

Heartwood: The older growth of the tree surrounding the pith.

Heat Capacity: The efficiency by which a material absorbs thermal energy.

Heat Resistance: The ability of the material to maintain its desired mechanical strength at elevated temperature.

Heat Treatment: A process of heating and cooling certain alloys to obtain desired properties.

Hemicellulose: A low-molecular-weight polymer formed from glucose.

High-Alloy Steel: Steels having a large amount of alloy elements added.

High-Carbon Steel: A plain carbon steel containing between 0.60 percent and 1.0 percent carbon.

Higher Fiber Volume Composite Concrete: Concrete containing 5-15% fiber.

Homopolymer: Polymer with identical monomers as repeating units in the polymer molecules.

Hot Isostatic Pressing (HIP): A forming process for composites that uses high temperatures and high pressure inert gas or liquid to densify and form composites.

Hot Pressing: A forming process that applies heat and mechanical pressure to ceramic powders in a die.

HSLA Steel: High-strength, low-alloy steel.

Hydration: Chemical process which causes the curing of concrete.

Hydraulic Cement: A cement that cures or hardens in water.

Hygroscopic: Readily absorbs and retains moisture.

Hysteresis Loop: An index of the lost energy in a complete cycle of magnetization.

I

Impact Test: A material test for impact resistance.

Impurities: Elements detrimental to material properties.

Index of Refraction: A ratio of light speed in a vacuum to light speed in a medium.

Induction Hardening: A surface hardening process using a magnetic field to rapidly heat the surface of a steel workpiece into austenitic condition; the piece is then quickly quenched.

Inert: Nonreactive elements that have reached the lowest energy level and, therefore, will not react with other elements.

Infiltration: A process of immersing fibers into a molten metal bath to produce a metal matrix composite.

Inhibitor: A chemical to prevent premature cross-linking in a thermoset polymer.

Initiator: A starting agent that reacts with polymer chains to produce cross-links for thermoset.

Injection Molding: A process where a material is forced into a mold of the desired product shape.

Insulators: Materials with low conductivity.

Interface: The boundary between the matrix and the reinforcement.

Intergranular Corrosion: A form of corrosion that occurs preferentially at the grain boundaries of a stainless steel.

Internal Oxidation: Oxygen dissolves in the metal, migrates, and forms oxide particles inside a material.

Interphase: The phase of the composite between the matrix and the reinforcement.

Ionic Bond: Atom-to-atom chemical bond in which one atom donates an electron to another, causing one to be negatively charged and the other to be positively charged.

Ionization: The process of removing or adding valence electrons to a neutrally charged atom.

Iron Ore: Natural mineral of iron oxides.

Iron-Iron Carbide Phase Diagram: A diagram showing the phases of iron and steel.

Isostatic Pressing: A forming process that uses a pressure vessel to apply fluid pressure to the outside of a mold containing powders.

Isotopes: Atoms of elements having only a few valence electrons from a cloud of electrons which are shared by metal atoms in order to reach the lowest energy level.

Isotropy: Exhibiting properties with the same values when measured along axes in all directions.

J

Jiggering: A forming process used for ceramics in which green clay is placed between matching mold halves. The clay and one of the mold halves is rotated causing the clay to conform to the shape of the mold.

K

Kaolin: Purest form of clay, often referred to as China clay.

L

Laminated Veneer Lumber (LVL): Made from specially graded veneers bonded together to form wider or longer composite lumber products.

Laser Beam Machining: A material removal process that uses focused light energy to separate or shape parts by material evaporation.

Laser Hardening: A surface hardening process using laser power to rapidly heat the surface of a steel workpiece, the heat is localized on the surface layer that is quickly quenched by the piece itself acting as a heat sink.

Late Wood (Summer Wood): The narrower, darker band of cells of the annual ring representing growth during the summer season.

Lay-Up: Simplest composite manufacturing process of building layer upon layer to obtain the desired thickness.

Lead Brass: Brass with a percentage of lead to improve machinability.

Lean Mixture: Proportionally less Portland cement in the concrete mixture than in a normal mix.

Lignin: A hard amorphous polymer that bonds cellulose and hemicellulose together to form a microfibril.

Linear Polymer: Polymer with a length much larger than the breadth of its molecular chain.

Liquid Carburizing: A process which uses salt to carry carbon into the surface layer, also known as salt carburizing.

Loading-Rate-Control Mode: Load is applied at a constant rate.

Low Fiber Volume Composite Concrete: Concrete containing less than 1 percent fiber.

Low-Carbon Steel: Steel containing less that 0.25 percent carbon.

Luminescence: Ability to give off visible light due to excitation by ultraviolet light.

M

Magnetism: A mutual attraction of two iron-based materials.

Mains Metal: Brass with a composition of 60 percent copper and 40 percent zinc.

Malleable Cast Iron: Cast iron with graphite precipitated from the solid matrix in the form of irregularly shaped nodules.

Malleablizing: A heat treatment process of making malleable iron.

Martensite: A very hard phase which provides hardening or strengthening on iron and steel.

Martensitic Stainless Steels: Stainless steels having martensitic structure at room temperature.

Matrix: The phase of a composite that is the glue or cement that bonds the reinforcement material.

Mechanical Property: A material's response to force.

Medium-Carbon Steel: A steel containing between 0.25 percent and 0.65 percent carbon.

Medullary Rays: A group of wood cells that grow in a radial direction from the center of the tree and transfer food from the roots through the pith.

Melamine: A thermoset plastic formed by reacting melamine with formaldehyde.

Metal Matrices: Matrices (see matrix) that are made from metals.

Metal Matrix Composites (MMC): Composites having metal or alloy as the major part, or matrix.

Microfibril: Submicroscopic elongated bundles of cellulose comprising the cell wall of a plant or tree.

Microhardness Tester: A hardness tester that performs a hardness test on the microstructure of a material.

Modulus of Elasticity: The ratio of stress to strain in elastic range for a material, also known as Young's modulus.

Molecular Weight: An index of molecular chain length, which is measured by the product of molecular weight of monomer multiplied by number of monomers in the polymer molecule.

Monomer: Basic repeating unit in a polymer molecule.

Monophase Ceramics: Ceramics composed of a single earth element.

N

Necking: The action of a material's cross section drastically reducing just prior to failure.

Network Polymer: Polymer of molecules that have cross-linked with each other and formed a network.

Neutrons: Subatomic particles in the nucleus of an atom that have no electrical charge.

Nitriding: A process in which nitrogen is diffused into the surface of the steel being treated.

Nodular Cast Iron: Cast iron with nodular or spherical graphite.

Noncrystalline Ceramics: Ceramic bodies having no orderly atomic structure; glass is an example.

Nondestructive Test: A material test that does not involve damage or destruction on the tested material.

Nonferrous Metals and Alloys: Refers to all the metals and alloys, except ferrous alloys, and are nonmagnetic.

Normal Stress: Stress perpendicular to the load on a material.

Normalizing: A process of heating the steel to a completely austenitic structure followed by air-cooling.

N-P-N Junction: A three-layer semiconductor consisting of a P type sandwiched between two N types; commonly used as a transistor.

N-Type: Doped semiconductors with donor atoms dominated by negative charge carriers.

O

Ohm's Law: Voltage (V) is a product of current (I) and resistance (R), V=IR.

Oriented Strand Board (OSB): Structured panel of wood strands specially oriented to produce maximum strength in all directions.

Outer Bark: The layer of the tree exposed to the environment that protects the other layers of the tree from weather and insects.

Oxidation: The reaction between a metal and oxygen without water or an aqueous phase.

Oxygen-Free Copper: Copper produced under a reducing atmosphere of carbon monoxide and nitrogen so that oxygen is prevented from entering the copper.

P

Pack Carburizing: A process in which a part is packed in a steel container so that the part is completely surrounded by granules of activated charcoal; carbon thus diffuses into the part.

Packing: The careful selection and mixing of ceramic powders to avoid porosity and large voids in ceramic composites.

Parallel Strand Lumber (PSL): Strands of wood or strips of veneer glued together under high pressure and temperature.

Parison: A hollow cylindrical tube of plastic melt during blow molding process.

Particulates: A classification of reinforcement material characterized by low aspect ratios.

Pauli's Exclusion: A law of physics stating that each quantum state of an atom can only be occupied by two electrons having opposite spin numbers.

Pearlite: A lamella structure of alternate plates of a ferrite and cementite (Fe_3C) which resembles mother-of-pearl.

Peltier Effect: When two different materials are in contact, the magnitude of the electromotive force (emf) across the junction is dependent upon the chemical composition of the metals and the temperature.

Penetrant Test: A nondestructive test using penetrant dyes.

Percent Elongation: A common measure of ductility in terms of amount of stretching relative to the original length before breaking.

Percent Reduction in Area: A common measure of ductility in terms of amount of reduction in cross sectional area relative to the original dimensions before breaking.

Periodic Table of Elements: Table of all chemical elements arranged according to atomic number.

Periods: Term used to refer to the rows in the periodic table of elements.

Permeability: The ratio of field strength to flux density of a material in a magnetic field.

Phenolics: Family of thermoset plastics known chemically as phenol formaldehyde.

Phosphorescence: A light-emitting material where the emission continues after ultraviolet excitation is removed.

Phosphorus Bronzes: Another name for tin bronzes. Tin bronzes frequently contain a noticeable amount of phosphorus (0.3%) from the refining process.

Photoelectric Effect: When a light is incident on certain material surface, electrons are emitted from the surface.

Photoelectrons: Electrons emitted from surface due to light excitation.

Pig Iron: Iron formed through a blast furnace, which has high carbon content.

Pith: Soft, sponge-like material at the very center of the tree trunk which transports and stores food for the other parts of the tree.

Pitting: A form of corrosion that develops in localized areas, resulting in cavities or pits.

Plain Carbon Steel: Alloys of iron and carbon.

Plain-Sawn Lumber: Logs that have been sawn into planks in a direction that is tangential to the annual rings.

Plastic Deformation: Deformation on material which cannot be recovered if the loading is removed.

Plasticity: A characteristic of a material which undergoes permanent deformation.

Plywood: Structural wood panel product manufactured by gluing thin veneers of logs together with the grain direction of adjacent veneers at right angles to one another.

P-N Junction: Bonding of a P-type and an N-type semiconductor; used primarily as a diode or a rectifier.

P-N-P: A three-layer semiconductor consisting of an N-type sandwiched between two P-type semiconductors; commonly used as a transistor.

Polar Moment of Inertia: A geometry property of a section, for a round section, polar moment of inertia is proportional to the fourth power of the diameter.

Polymer Matrices: Matrices (see matrix) that are made from polymers.

Polymer: A matter with molecules made up of many repeating units (monomers).

Polymerization: A process of linking repeating molecules (monomers) together to form a polymer.

Polymorph: Means "many forms" and refers to materials that can change their structure with changes in their environment.

Polyolefin Family: A family of thermoplastics having similar molecular structure to polyethylene.

Polyphase Ceramics: Ceramics composed of more than one earth element.

Polypropylene: A polymer with propylene monomers.

Polystyrene: A polymer made of styrene monomers.

Porcelain Enamels: Glassy ceramic coatings applied to kitchen and bathroom appliances and fixtures.

Portland Cement: Finely pulverized material consisting of compounds of lime, silica, alumina, and iron.

Pot Life: The duration for resin mixed with initiator to be stored without significant hardening.

Powder Metallurgical Process: A process of producing metal matrix composite from powders.

Powder Pressing: The selection, mixing, and forming of ceramic powders into a single composite part or product.

Pozzolana: Volcanic ash from Mount Vesuvius.

Preform: An array of tooling that squeezes away excess resin.

Prepreg: A form of fibers that are preimpregnated with resin.

Protons: Positively-charged subatomic particles in an atom's nucleus.

P-Type: Doped semiconductors with donor atoms dominated by positive charge carriers.

Puddling: A term used to describe the prescribed action of raising and lowering a 5/8″ bullet-nosed steel rod in specimens of concrete used for slump and compression testing.

Pultrusion: A composite manufacturing process for producing continuous lengths of reinforced composite structural shapes.

Pyrolysis: The burning of a polymer, such as rayon, in a high-temperature furnace to produce a strong, stiff fiber.

Q

Quarter-Sawn Lumber: Logs that have been sawn into planks in a radial direction to the center of the tree.

Quenching: A process involving heating a steel to high temperature followed by rapidly cooling after holding it at high temperature for some time.

R

Radiographic Test: A nondestructive test using radiation, such as x ray.

Reaction Injection Molding (RIM): A forming process used for polyurethane foams in which polymer reactants are pumped from separate reservoirs into a mixing chamber and then directly to a heated mold where the product is formed and cured.

Rebar: Reinforcement bar used in concrete to increase its flexural strength.

Reconstituted Wood Products: Structural and nonstructural products made from smaller components of the tree with the use of adhesives, heat, and pressure.

Red Brass: Brass that contains about 15% zinc.

Reflectivity: The ability of a material to produce a specular reflection.

Refraction: Bending of a light ray.

Refractory Ceramics: High-melting-point oxide ceramics frequently referred to as crystalline ceramics.

Reinforcement: Additive with large aspect ratio, normally able to improve the strength of polymer composite.

Relative Permittivity: The charge-storing capacity of a dielectric relative to a vacuum.

Remnant Induction: The magnetic flux remaining in the material after the magnetic field is removed, as in the permanent magnet.

Residual Strain: Deformation or strain remaining on a material after the load has been removed.

Resin System: A combination of resin, curing agents, fillers, and inhibitors.

Resin Transfer Molding (RTM): A process in which an essentially monomeric material is injected into the mold; then resin is injected and reacts to form thermoset polymer of finished products.

Resistivity: A materials constant which is related to the electrical resistance per unit length per unit cross sectional area.

Rich Mixture: Proportionally more Portland cement in the concrete mix than in a normal mix.

Rockwell Hardness Test: A material hardness test based on the depth of penetration by an indenter.

Rockwell Superficial Hardness Test: A Rockwell hardness test using less load for testing the hardness of soft or sheet materials.

Rule of Mixtures: The property of a composite is directly related to the property and fraction of each constituent.

S

Salt Carburizing: Also known as liquid carburizing.

Sapwood: Newer growth of the tree consisting of living wood cells that transport moisture to the leaves from the roots.

Scale: A thick surface oxide film developed when a metal or alloy is exposed to high temperature.

Sea of Electrons: Also referred to as electron cloud; accumulation of electrons in metals that forms during chemical bonding.

Seebeck Effect: The total electromotive force of a thermocouple depends on the temperature difference between the two junctions.

Semicrystalline Polymer: A polymer that has a combination of amorphous and crystalline structure.

Set: A term used to describe the curing of concrete into a hard mass.

S-Glass: A type of glass fiber used in fiberglass composites when high strength is needed.

Shear Modulus of Elasticity: A proportional constant or ratio of shear stress to shear strain.

Shear Strain: The angular change of the element caused by shear force.

Shear Strength: The maximum shear stress a material can withstand.

Shear Stress: The intensity of the force parallel to the surface of a material.

Sheet Molding Compound (SMC): A continuous sheet of ready-to-mold composite material containing fibers and mineral fillers dispersed in a thermoset resin.

Shelf Life: Also known as storage life; the duration for which a resin can be stored.

Silicates: Compounds of silicon and oxygen whose structure has an integral arrangement of tetrahedral units.

Silicon Bronze: A no-tin copper alloy that contains 1% to 3% silicon.

Silicones: Polymers composed of monomers in which oxygen atoms are attached to silicon atoms.

Sintering: Applying heat to a ceramic body causing viscous flow of powders to reduce porosity and increase densification.

Slaked Lime: Lime heated and crumbled by adding water.

Slip Casting: A forming method used to produce ceramic products with hollow shapes, e.g., toilets, sinks.

Slip Clay: A low purity clay used as a glaze.

Slump Test: A standardized test used to determine the consistency of concrete.

Slurry Casting: A process in which liquid metal is simply stir-mixed with solid ceramic particles and the mixture is allowed to solidify on cooling.

S-N Diagram: An endurance curve.

Soft Magnet: Material that is easily magnetized and demagnetized.

Solution Treatment: A process of heating an aluminum alloy to a high temperature and then quickly quenching the alloy.

Specular Reflection: Reflection of light from a smooth surface.

Spherulitic Graphite Iron: Also known as nodular cast iron.

Splitting Tension Test: A test procedure of testing tensile strength of concrete by compressing a cylindrical sample along its diameter.

Squeeze Casting: A process in which pressure is imposed on a solidifying alloy, usually by a single hydraulically activated ram.

Squeeze Infiltration: A process in which liquid metal is injected into interstices of an assembly of long fibers.

Stainless Steel: Steel of high corrosion resistance, usually having a large amount of Cr and Ni.

Static Loads: Any load applied slowly in one direction.

Steel: Ferrous alloy containing less than 1.7 percent of carbon.

Storage Life: The duration of time for which a resin can be stored.

Strain: The amount of deformation on a material per unit length.

Strain Hardening: The increase in hardness or strength of a material that has undergone plastic deformation.

Strain-Rate-Control Mode: Rate of specimen elongation (strain) is constant.

Stress Corrosion Cracking: A form of corrosion that is accelerated by a stress and a specific chemical.

Stress Relaxation: The stress required to maintain the same amount of strain decreases with time.

Striking Off: Removing excess concrete by carefully moving a metal bar back and forth across the top of a form or testing vessel while pulling the bar forward.

Structural Clay: Ceramics made from abundant low cost clays; structural clay products are used where strength is required, e.g., concrete blocks, foundation walls, drainage tile.

Structural Composite Lumber (SCL): Structural products made by gluing veneers and strands of wood together to form a single structural member.

Styrene: A monomer formed by substituting one hydrogen with benzene ring in ethylene.

Superalloys: Alloys containing more than 50 percent of alloying elements.

Superconductor: A perfect conductor which has no resistance to the flow of electrons.

Superplastic Forming: A process in which a material can be deformed almost indefinitely under suitable conditions without local fracturing.

T

Tape Casting: Process involving the drawing of a powdered slurry under a blade forming a thin layer of ceramic on a moving plastic belt. Evaporation causes the moisture to be removed forming a ceramic tape.

Tempering: A heat treatment process in which the quenched parts are reheated to a temperature much lower than quenching temperature, held at this temperature for a relatively long period, and cooled at any desired rate.

Tensile Test: A test procedure of gradually applying tension on specimen until fracture occurs.

Thermal Conductivity: Ability of a material to transfer heat from a body of higher temperature to one of lower temperature.

Thermal Expansion: The degree to which a material increases in volume with an increase in temperature.

Thermal Properties: The response of materials to changes in temperature of the environment.

Thermoplastics: Polymers that soften upon heating and regain their solid form when cooled again.

Thermoset Plastics: Polymers that can be passed through only one heat cycle; and cannot be re-formed by thermal process.

Thomson Effect: When there is a temperature difference between the ends of a single homogeneous wire, the electromotive force (emf) is dependent upon the composition, the chemical uniformity of the material, and the temperature difference.

Tin-Bronze: A copper-tin alloy.

Tool Steel: High alloy steels used to make tools or dies.

Torsion: A torque (twist) shear stress.

Transfer Molding: A type of compression molding that uses a double chamber to minimize the direct pressure on the die cavity.

Transition Temperature: A critical temperature for materials, above which materials are ductile; below which materials are brittle.

U

Ultimate Tensile Strength: The maximum stress which can be applied to a material before it fails.

Ultrasonic Test: A nondestructive test utilizing ultrasound.

Uniform Attack: A chemical or electrochemical reaction that proceeds uniformly over the entire exposed surface, also known as general corrosion.

Unit Cell: Repeating patterns of three dimensional structures in crystalline solids.

Unsaturated Polyesters: A large group of synthetic resins produced by condensation of acids with an alcohol or glycol.

Urea Formaldehyde: A synthetic thermoset polymer made by reacting amino resin with urea and formaldehyde.

Urethanes: Group of polymers based on polyester or polyether resins.

V

Vacuum Bagging: A process that applies vacuum to help compress composite plies and withdraw volatile materials.

Valence Shell: The outermost orbit or energy level of an atom.

Van der Waals Forces: Weaker forces of attraction producing bonds between atoms or molecules.

Vinyl Chloride: A monomer formed as a result of one carbon in ethylene substituted with chlorine (Cl).

Viscoelastic Behavior: Mechanical behavior of polymer having certain ingredients of both elastic solid and viscous fluid.

Viscoelasticity: A delayed material response to stress after the stress is applied and kept constant on the polymer.

Vitreous: Usually a reference to glass.

Vitrification: A process which uses high temperature to cause partial melting of ceramic bodies to produce a hard, dense structure.

W

Wafer Board: A nonveneer structural panel made by gluing wood flakes together under pressure.

Weathering Steel: A copper-bearing, low-alloy steel.

Wet Filament Winding: Also known as filament winding.

Wet Process: One of two processes used to manufacture Portland cement. During this process the properly proportioned raw materials are ground with water, mixed, and fed into a large rotary kiln.

Wet Winding: Also known as filament winding.

Wetability: The ability of a liquid material to envelope, or coat, the dry constituents in a composite.

Whiskers: Very fine single-crystal fibers that range from 3-10μm in diameter.

White Cast Iron: Cast iron in which carbon exists in the form of iron carbide or cementite.

Y

Yard Lumber: A classification of softwood lumber used primarily for casework, windows, and doors.

Yellow Brass: Brass with 30% zinc.

Yield: A condition where considerable elongation or deformation occurring on a material.

Yield Point: A point on the stress-strain curve which shows the occurrence of yield.

The stress level where a material begins to plastically deform.

METRIC CONVERSION TABLE

	To Metric			From Metric		
	To Convert From	To	Multiply By	To Convert From	To	Multiply By
Length	inches (in)	millimeters (mm)	25.4	millimeters (mm)	inches (in)	0.03937
	feet (ft)	meters (m)	0.3048	meters (m)	feet (ft)	3.2808
Mass	ounces (oz)	grams (g)	28.3495	grams (g)	ounces (oz)	0.03527
	pounds (lb)	grams (g)	453.59	grams (g)	pounds (lb)	0.002203
	pounds (lb)	kilogram (kg)	0.4536	kilogram (kg)	pounds (lb)	2.2046
	pounds (lb)	metric tons	0.000454	metric tons	pounds (lb)	2205
	tons	metric tons	0.907	metric tons	tons	1.1025
	metric tons	kilograms (kg)	1000	kilograms (kg)	metric tons	0.001
Pressure/Stress (Strength)	pounds/square inch (psi)	ksi (1000 psi)	0.001	ksi (1000 psi)	pounds/square inch (psi)	1000
	ksi	megapascals (MPa)	6.895	megapascals (MPa)	ksi	0.145033
	pounds/square inch (psi)	megapascals (MPa)	0.006895	megapascals (MPa)	pounds/square inch (psi)	145.033
	newtons/sq mm (N/mm^2)	megapascals (MPa)	1	megapascals (MPa)	newtons/sq mm (N/mm^2)	1
	pounds/square inch (psi)	newtons/sq mm (N/mm^2)	0.006895	newtons/sq mm (N/mm^2)	pounds/square inch (psi)	145.033
	ksi	newtons/sq mm (N/mm^2)	6.895	newtons/sq mm (N/mm^2)	ksi	0.145033
	ksi	kilogram/sq millimeter (kg/mm^2)	0.704	kilogram/sq millimeter (kg/mm^2)	ksi	1.42045

Appendix B

	PROPERTIES OF SELECTED ELEMENTS					
Element	**Atomic Number**	**Symbol**	**Atomic Mass (g/mol)**	**Density (g/cm³)**	**Melting Point (°C)**	**Crystal Structure**
Aluminum	13	Al	26.91	2.699	660.2	FCC
Antimony	51	Sb	121.757	6.62	630.5	Rhombo
Arsenic	33	As	74.921	5.72	817	Rhombo
Beryllium	4	Be	9.012	1.848	1277	HCP
Bismuth	83	Bi	208.980	9.80	271.3	Rhombo
Boron	5	B	10.811	2.30	2300	Rhombo
Cadmium	48	Cd	112.411	8.65	320.9	HCP
Calcium	20	Ca	40.078	1.55	838	FCC
Carbon	6	C (graphite)	12.011	1.50–1.80	3826	Hex
Cerium	58	Ce	140.115	6.67	799	FCC
Cesium	55	Cs	132.905	1.903	28.7	BCC
Chromium	24	Cr	51.96	7.19	1875	BCC
Cobalt	27	Co	58.933	8.85	1495	HCP
Copper	29	Cu	63.546	8.96	1083	FCC
Fluorine	9	F	19.0	1.3	−220	—
Gallium	31	Ga	69.723	5.907	29.78	Orthorhombic
Germanium	32	Ge	72.61	5.323	937.4	Diamond Cubic
Gold	79	Au	196.967	19.32	1063	FCC
Hafnium	72	Hf	178.49	13.09	2222	HCP
Indium	49	In	114.82	7.31	156.2	FC Tetragonal
Iron	26	Fe	55.847	7.874	1536.5	BCC
Lead	82	Pb	207.2	11.34	325.6	FCC
Lithium	3	Li	6.941	0.534	180.54	HCP
Magnesium	12	Mg	24.305	1.738	650	HCP
Manganese	25	Mn	54.938	7.43	1245	Cubic
Mercury	80	Hg	200.59	13.546	−38.9	Rhombo
Molybdenum	42	Mo	95.94	10.22	2610	BCC
Nickel	28	Ni	58.693	8.902	1453	FCC
Niobium	41	Nb	92.91	8.6	2468	BCC
Osmium	76	Os	190.2	22.583	2700	HCP
Oxygen	8	O	15.94	1.43	−218	Cubic
Palladium	46	Pd	106.42	12.02	1552	FCC
Phosphorus	15	P	30.974	1.84	44.2	Cubic
Platinum	78	Pt	195.08	21.45	1769	FCC
Potassium	19	K	39.098	0.86	63.7	BCC
Rhenium	75	Re	186.207	21.04	3180	HCP
Rubidium	37	Rb	85.468	1.532	38.9	BCC
Ruthenium	44	Ru	101.07	12.2	2500	HCP
Selenium	34	Se	78.96	4.79	217	Hex
Silicon	14	Si	28.085	2.33	1410	Diamond Cubic
Silver	47	Ag	107.868	10.49	960.8	FCC
Sodium	11	Na	22.990	0.97	97.8	BCC
Tantalum	73	Ta	180.948	16.6	2996	BCC
Tin	50	Sn	118.710	7.30	231.9	Diamond Cubic
Titanium	22	Ti	47.88	4.507	1668	HCP
Tungsten	74	W	183.85	19.3	3410	BCC
Vanadium	23	V	50.91	6.1	1900	BCC
Yttrium	39	Y	88.906	4.47	1509	HCP
Zinc	30	Zn	65.39	7.133	420	HCP
Zirconium	40	Zr	91.224	6.489	1852	HCP

GENERAL COMPARISON OF PROPERTIES: METALS, CERAMICS, POLYMERS			
Property	**Metals**	**Ceramics**	**Polymers**
Density	2-22 (ave. ~8)	2-19 (ave. ~4)	1 to 2
Melting points, °C	Low = gallium 30; High = Tungsten 3410	High 4000	Max. serv. temp. ranges from ~45 for ethyl cellulose to ~300 for polyimide
Hardness	Medium	High	Low
Machinability	Good	Poor	Good
Tensil Strength, MPa (ksi)	Up to 2500 (360)	Up to 400 (58)	Up to 140 (20)
Compressive Strength MPa (ksi)	—	Up to 5000 (725)	Up to 350 (50)
Young's Modulus GPa (10^6 psi)	15 to 400 (2 to 58)	150–450 (22–65)	0.1 to 10 (0.00015 to 1.45)
High Temp. Dreep Resistance	Poor to Mediium	Excellent	—
Thermal Expansion	Medium to High	Low to Medium	Very High
Thermal Conductivity	Medium to High	Medium, but often decreases with temp.	Very Low
Thermal Shock Resistance	Good	Generally Poor	—
Electrical Properties	Conductors	Insulators	Insulators
Chemical Resistance	Low to Medium	Excellent	Good
Oxidation Resistance	Generally Poor	Oxides excellent SiC and Si_3N_4 Good	—

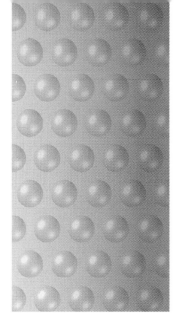

Index